MW00985080

RISK ASSESSMENT AND RISK MANAGEMENT FOR THE CHEMICAL PROCESS INDUSTRY

RISK ASSESSMENT AND RISK MANAGEMENT FOR THE CHEMICAL PROCESS INDUSTRY

Stone & Webster Engineering Corporation

Edited by

Harris R. Greenberg
Joseph J. Cramer

VNR VAN NOSTRAND REINHOLD
New York

Copyright © 1991 by Van Nostrand Reinhold

Library of Congress Catalog Card Number 90-22775
ISBN 0-442-23438-4

Printed in the United States of America

Van Nostrand Reinhold
115 Fifth Avenue
New York, New York 10003

Chapman and Hall
2-6 Boundary Row
London, SE1 8HN, England

Thomas Nelson Australia
102 Dodds Street
South Melbourne 3205
Victoria, Australia

Nelson Canada
1120 Birchmount Road
Scarborough, Ontario MIK 5G4, Canada

16   15   14   13   12   11   10   9   8   7   6   5   4   3   2

Library of Congress Cataloging-in-Publication Data

Risk assessment and risk management for the chemical process industry/Stone & Webster Engineering Corporation; edited by Harris R. Greenberg, Joseph J. Cramer.
       p.   cm.
   Includes index.
   ISBN 0-442-23438-4
     1. Chemical plants—Risk assessment. Risk management. I. Greenberg, Harris.
II. Cramer, Joseph.   III. Stone & Webster Engineering Corporation.
TP155.5.R55   1991
660'.2804—dc20   90-22775
                                                                      CIP

Contents

	Preface	Joseph J. Cramer Harris R. Greenberg	vii
	Introduction	Harris R. Greenberg	ix
	Contributors		xvii
1	Risk Management Programs	Robert W. Myers Joseph J. Cramer Robert T. Hessian, Jr.	1
2	Screening Analysis Techniques	Michael E. Sawyer Michael C. Livingston William F. Early	15
3	Checklist Reviews	Robert T. Hessian, Jr. Jack N. Rubin	30
4	Preliminary Hazards Analysis	Robert T. Hessian, Jr. Jack N. Rubin	48
5	Safety Audits	Robert T. Hessian, Jr. Jack N. Rubin	57
6	WHAT-IF Analysis	William W. Doerr	75
7	Failure Modes and Effects Analysis	Robert L. O'Mara	91
8	Hazard and Operability Studies	Robert M. Sherrod William F. Early	101
9	Fault Tree and Event Tree Analysis	Harris R. Greenberg Barbara B. Salter	127
10	Chemical Plume Dispersion Analysis	Stephen A. Vigeant	167

11   Explosion and Fire   Frank A. Elia, Jr.   196
            Analysis

12   Assessment of Health   Barbara B. Waterhouse   209
            Effects from Chemical   Virginia A. Cava
            Releases   Victoria M. Cathcart

13   Quantified Risk   Robert L. O'Mara   221
            Assessment   Harris R. Greenberg
                                          Robert T. Hessian, Jr.

14   Calculation of Human   Robert L. O'Mara   238
            Reliability

15   Training for Industrial   Robert W. Myers   253
            Facilities   Dixie J. Finley

16   Emergency   Carolyn C. Burns   263
            Preparedness   Piero M. Armenante

17   Risk Financing   Michael J. Natale   305

18   Computer Techniques   Kenneth F. Reinschmidt   316

19   Directions in   Joseph J. Cramer   343
            Legislation and
            Regulation

Acronyms   353

Index   356

Preface

The chemical industry, with its consultants and engineers, has been concerned with chemical safety from the industry's earliest days. Techniques have evolved from the two-legged stool used during the early preparation of nitroglycerine to include the application of probabilistic risk assessment and expert systems. Although many safety concepts are not new, it is true that much has happened to crystallize and focus industry's thinking in the six years since the tragic accident at Bhopal, India. Industry groups like the AIChE's new Center for Chemical Process Safety, the Chemical Manufacturers Association, the American Petroleum Institute, and the National Safety Council, to name several, have attempted to organize, prioritize, and extend the techniques for hazard management, analysis, and mitigation commonly used by the leading practitioners. Regulatory agencies on both the state and federal levels have also begun to shape new directions for hazards management thinking.

The nineteen chapters contained in this book deal with the detection, prevention, and mitigation of the risks associated with the processing, handling, and production of hazardous chemicals. Techniques reviewed range from the simplest experience-based checklist approaches to the use of advanced risk assessment and expert system analyses. The use of various insurance vehicles to deal with residual risk and a review of recent regulatory developments are also addressed to make the treatment of process hazards management more comprehensive. The authors are nearly all senior staff members of Stone & Webster Engineering Corporation who have been actively involved in the application of the various techniques and methodologies described in the individual chapters. The disciplines represented by the authors include nearly all the standard engineering branches as well as meteorology, health sciences, computer science, insurance, management consulting, and education. The theoretical insights contained within have been strongly flavored as a result of the authors' many years of collective practical application.

The order of the chapters generally proceeds from the broad-based overview of risk management or process hazards management through increasingly detailed hazard identification and risk assessment methodologies. Chapters dealing with the use of equipment failure rate and human reliability data, risk financing, new computer tools, and evolving regulatory directions are near the end of the book. To keep reader appeal as broad as possible, control technology and specific operations or maintenance practices are not treated in detail. Other treatises dealing primarily with specific types of controls, treatment, and practices are available or in preparation by others. Several projects presently being undertaken by the Center for Chemical Process Safety deal with these specifics. The focus of this work is the integration of hazard identification and assessment and consequence analysis and mitigation into a comprehensive formalized program to manage the risks associated with the handling of hazardous chemicals.

Although an issue such as chemical safety, which is strongly affected by real-life events, political developments, economic trends, and public pressures, is continuously evolving and changing, every effort was made to present the material in a general enough fashion to delay the onset of obsolescence. Many of the approaches described in the greatest detail have been utilized in industry for some years and are likely to continue in use for the foreseeable future.

Many of the references at the end of each chapter have also stood up well to the tests of time and should be useful to those desiring to make a more detailed study of any particular aspect of an overall process hazards management program.

We would like to thank Stone & Webster management for supporting this work and the many authors of the individual chapters for their extensive efforts, much of it on their own time. We especially want to thank Barbara Moorhouse, Virginia Cava, and Arthur Schwartz, staff editors at Stone & Webster, for ably bearing the heavy burden involved in handling the editorial responsibilities necessary to complete this book.

Introduction

This book is intended as an aid to the practicing engineer or manager who has the responsibility for risk assessment, risk management, and risk reduction for a complex piece of equipment, system, or facility.

Getting past the jargon is often half the battle when working in a new field, and we will attempt to minimize the use of jargon. However, since this book is not likely to be the only source you will be using, we have included a list of acronyms in Appendix A.

As we attempt to define risk and risk assessment, we assume that the reader has some general idea about what these terms imply, and we discuss the overall scope of the subject. There are many elements to a program to manage risk at a given facility. Because there is no such thing as zero risk, the best we can do is to minimize risk, keeping any existing ones under control and not introducing any new ones.

Figure 1 shows the various elements of a risk management program. This overall view of risk management and risk reduction plans is discussed in detail in Chapter 1. This book addresses most of the individual elements of the risk management program pie, some with more detail and emphasis than others. The specific techniques for incident and accident investigation are not discussed separately, since they are variations on the techniques

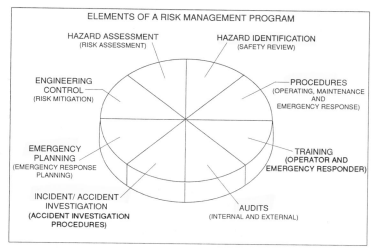

Figure 1   Elements of a risk management program.

used to identify and assess potential accidents before they happen, and those we cover in great detail.

Specific engineering control and mitigation measures are touched on in the examples discussed throughout the book, but since they are often specific to a given hazardous material, there is not sufficient space to cover this subject adequately here. A computerized literature search of any of a number of on-line data bases will rapidly identify chemical-specific engineering control and mitigation options. The techniques in the text can then be applied to assess the potential benefit of the various options identified from the computer search.

The two elements of risk management that get the most extensive treatment are hazard identification and hazard assessment. Hazard identification techniques are discussed in Chapters 2 through 8. The hazard identification techniques range from fairly simple and straightforward screening analysis (Chapter 2) and checklists (Chapter 3) to highly structured and systematic review techniques such as Hazard and Operability (HAZOP) analysis (Chapter 8).

Hazard assessment techniques are discussed in Chapters 9 through 14. Fault Tree and Event Tree analyses, covered in Chapter 9, are used to determine how undesirable events can occur. Chapters 10, 11, and 12 provide information on calculating the potential consequences of the hazards identified using the techniques discussed in Chapters 2 through 8. Chapters 13 and 14 deal with quantifying the likelihood of these undesirable events, combining equipment failure rate and human reliability analysis with the techniques discussed in Chapter 9.

Chapter 15 deals with both operator training and training of emergency response personnel. Chapters 16 and 17 discuss other elements of risk management including emergency preparedness, risk communication, and risk financing and transference.

Rounding out our treatment of risk assessment and risk management, Chapter 18 deals with the use of artificial intelligence and computer expert system techniques as applied to risk assessment and risk management, and Chapter 19 discusses future outlooks for safety legislation and regulation.

DEFINITIONS OF RISK AND RISK PERCEPTION

The formal definition of risk is

> The combination of probability that an undesired event will occur and the consequences that occur as a result of the undesired event.

i.e., RISK = PROBABILITY × CONSEQUENCES

The following list compares risk perceptions of the public to risk perceptions by "technical experts."

PUBLIC	EXPERT
Subjective	Objective
Qualitative	Quantitative
Uninformed	Knowledgeable
Irrational	Rational
Intuitive	Logical
Misperception	Reality

In applying a risk definition like the preceding one, the technical expert is trying to be objective and quantitative. He or she attempts to apply logic by dealing knowledgeably with facts and reality. Risks can be perceived very differently by the public and by experts. The public regards risks subjectively and qualitatively, from the individual standpoint, and often reacts from gut feeling rather than from rational thought. Very often under these conditions, the "risk" is perceived only as the potential consequence without regard to the likelihood of occurrence. There are times when even the "experts" fall prey to their emotions, especially when there is a potential for direct impact on themselves or their families.

The following list presents another view of how society perceives risk and what constitutes acceptable risks. These are general observations, but they help define specific attributes of risk perception which can sometimes be modified, thereby modifying the public acceptability of certain risks.

ACCEPTABLE RISKS	UNACCEPTABLE RISKS
Known	Unknown
Old	New
Gradual	Sudden
Usual	Unusual
Natural	Manmade
Voluntary	Involuntary
Controllable by individual	Uncontrollable by individual
Necessary	Luxury
Profitable for individual	Not profitable for individual or profitable for a company
Entertaining/Recreational	No entertainment value

Risks that are known are more acceptable than those that are unknown. A good example is the risk of smoking cigarettes, which is a known risk, versus the risk of breathing the strange-smelling gas that comes from a chemical process plant. Breathing the gas might be less dangerous than breathing the cigarette smoke, but it is perceived as a less acceptable risk.

Old and familiar risks are more acceptable than new risks. An example is the acceptable risk from an existing facility that has been around a long time, compared to the unacceptable risk from the same plant or a similar plant being built today, perhaps in the same neighborhood. The older facility may have out-of-date or aging equipment, and the new facility may have all of the state-of-the-art safety features that money can buy, but the new one is rejected as posing an unacceptable risk simply because it is a risk that was not there before.

An example of the difference in risk perception for "gradual" versus "sudden" is the risk of traveling by car versus the risk of traveling by airplane. In car accidents, usually only one or two people are killed or injured at a time, while in an airplane crash hundreds of people may be killed or injured at one time. The likelihood of being killed in a plane crash while flying from New York to Washington is much lower than for driving the same distance by car; however, more people fear the plane ride than the automobile ride. Many more people die in automobile accidents every year in the United States than die in plane crashes, but the sudden deaths of hundreds of people make a more lasting impression.

Let us compare the "usual" risk of getting indigestion from eating a greasy pepperoni pizza and the "unusual" risk of getting indigestion from a piece of rare filet mignon you eat at a fine restaurant. Although you may be in more danger from eating your steak rare, you expect the possibility of indigestion from the greasy pizza, so it is a more acceptable risk than the case of indigestion you get from the undercooked steak. You are more likely to complain bitterly about the fine restaurant than you are to complain about the pizza parlor.

Risks that are natural, or acts of God, are more readily acceptable than risks imposed by things artificially constructed. A particular community might lose millions of dollars from water damage when a river overflows its banks, but this risk is accepted by the people who choose to live in that community. However, the same community residents would not accept being told that their houses would be flooded periodically because of water main breaks. Even though the damage from the flooding river is likely to be much more extensive and may occur more frequently than water main breaks, the water main break is from an artificial source.

Risks that are voluntary, controllable by the individual, or considered necessary are more acceptable than risks that are involuntary, not controllable by the individual, or considered to be a luxury rather than a necessity. A person might voluntarily cross a stream on a rickety log, because it was something she wanted to do, or because it was necessary to get to the other side of the stream. However, risks present in the workplace that might be less dangerous would not be accepted because the person is required by others to be there, and the person may not have any control over whether she is exposed to the risk.

Risks perceived to benefit the individual are more acceptable to the individual than risks not perceived as beneficial. For example, skydiving or skiing may be very risky activities, but people do both because they enjoy the sports.

Understanding all these factors can be useful in a practical sense, especially in dealing with the community or the press when a new project is proposed. Having a well-designed, modern facility with very low risk to plant workers or the public is not enough to gain acceptance; the facility must also be perceived as safe by both the workers and the community.

SAFETY ENGINEERING: YESTERDAY AND TODAY

Human beings learned to build safely by trial and error, learning from their mistakes and passing down *rules* from lessons learned the hard way. Then, as engineering science matured, people were able to design simple structures and systems based on performance requirements. Safety was built into designs in the form of a margin beyond the calculated requirements. If a bridge needed to support a certain weight as a worst case, it was designed to take twice that load, and so it had a safety factor built in.

However, safety margins alone do not ensure the safe usage of a piece of equipment, a structure, or a system. How something can be misused is often as important as how it is designed. Human error, poor maintenance after installation, and interactions of a system or piece of equipment with the environment or with other systems can be overlooked in the design process with disastrous effects.

The methods of hazard identification and risk assessment discussed in this book are practical methods used today in industry. Some are simple and straightforward, and some are highly structured and systematic. Rigorous and systematic methods are required when dealing with highly complex systems or systems with high potential risks, to ensure that all possible dangers are identified and dealt with.

The techniques presented here can be used to deal with risks that are very unlikely but potentially catastrophic. Traditional system design might have considered "double jeopardy" beyond design requirements, but history has shown that the more disastrous events often have had multiple contributing failures—human as well as mechanical. As systems become larger or as populations grow around a once-isolated plant, the potential for disaster and consequences increases, and more possibilities need to be investigated to ensure that the risk to workers and the public is being dealt with acceptably.

INHERENTLY SAFE VERSUS EXTRINSICALLY SAFE

Safety can be considered at different stages in the life of a component, system, or facility. The earlier in the process it is considered, the less it costs to implement and the greater the opportunity to have an inherently safe design. By *inherently safe,* we mean that potential dangers have been removed rather than designed for.

Consider the production of a particular product that requires a highly toxic or explosive chemical to be stored on site. A modification that would make this process inherently safe would be a change in the process that eliminated the need to store the dangerous chemical on site. This could be done in several ways, such as changing the process so that the dangerous chemical is not used, is not stored in quantity on site, or is used up as fast as it is produced in another part of the plant.

The life of a plant can be divided into five stages: process development,

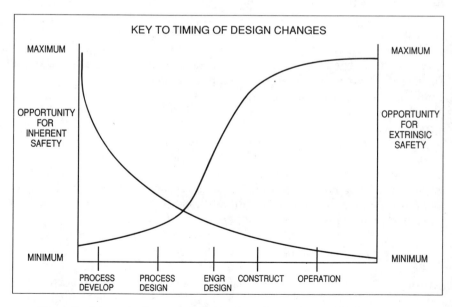

Figure 2   Inherent safety vs. extrinsic safety. Key to timing of design changes.

process design, engineering design, construction, and operation. Later in the design, safety considerations are manifest as extrinsic safety. That is, the safety is built in by adding controls, alarms, interlocks, equipment redundancy, safety procedures, and other mechanisms. At the later stages of plant life, there are still opportunities for extrinsic safety, but very little opportunity for inherent safety. This is shown in Figure 2.

Contributors

Joseph J. Cramer received B.S., M.S., and Ph.D. degrees in chemical engineering from the University of Pennsylvania and Massachusetts Institute of Technology. After obtaining his doctorate in 1971 from Penn, Dr. Cramer began with Stone & Webster Engineering Corporation, first as an environmental engineer and then as a project manager and program manager. For five years, Dr. Cramer acted as program manager for Stone & Webster's Process and Chemical Plant Safety Program. He has coordinated and directed all technical, developmental, and marketing activities dealing with process safety at all Stone & Webster offices. Dr. Cramer has been a member of AIChE for nearly 20 years and a member of the Environmental Division for some time. He is currently chair of AIChE's Environmental Division.

Harris R. Greenberg, P.E., is the manager of Plant Safety Programs at Stone & Webster Environmental Services Division and assistant manager of the Nuclear Projects Division of Stone & Webster Engineering Corp. With over 21 years of engineering and design experience, he has directed several chemical and petrochemical plant risk assessments, including both probability and consequence calculations for toxic plume dispersion, explosions, and fires. For the past four years he has taught Stone & Webster's risk assessment and hazard analysis courses. Mr. Greenberg is currently the chairman of the ASME Safety Division's Facility Safety Committee, and member-at-large of the ASME Safety Division's Executive Committee. Mr. Greenberg has a B.S. degree in physics from Cooper Union, a masters of science degree in nuclear engineering from Massachusetts Institute of Technology, a degree in chemical engineering from MIT, and a certificate in management from Polytechnic Institute of New York.

Piero M. Armenante received his M.S. degree from the University of Rome, Italy, and his Ph.D. from the University of Virginia, both in the chemical engineering. He has worked in both industry and academia, and has served as a consultant for international organizations such as the United Nations Industrial Development Organization and the International Institute for Applied Systems Analysis. In 1984, Dr. Armenante joined the faculty of the New Jersey Institute of Technology, Department of Chemical Engineering, Chemistry and Environmental Science. He has authored more than 35 papers in his field, and his current research activities involve treatment of hazardous and toxic waste, plant safety, and biochemical processes modeling.

Carolyn C. Burns is a principal scientist in the emergency preparedness group at Stone & Webster Engineering Corp., and a registered environmental assessor in the state of California. She earned her A.B. in biochemistry from Vassar and an M.S. in chemistry at Simmons. Ms. Burns holds two graduate certificates from Harvard, one in natural sciences and one in administration and management. She has prepared and evaluated emergency preparedness programs for domestic and international manufacturing and power facilities. She is a coinstructor of the American Institute of Chemical Engineers course on emergency response planning and has had extensive teaching experience in chemistry.

Victoria M. Cathcart is a chemical engineer and an industrial hygienist. She has had experience in industrial hygiene and manufacturing engineering and is currently responsible for performing industrial hygiene evaluations to assess possible health effects on workers and the community. She has administered training and monitoring programs, including right-to-know and respiratory protection, and she has conducted industrial hygiene audits.

Virginia A. Cava, P.E., is a mechanical engineer with eight years' experience in mechanical system design and analysis, including power generation, noise suppression and human engineering. She has been involved in several Stone & Webster process risk assessment projects, including chemical release consequence analyses, and various environmental impact studies for cogeneration, waste treatment, and resource recovery facilities. Ms. Cava is a licensed professional engineer in the state of New York, and a member of ASME and SWE.

William W. Doerr, Ph.D., has more than 21 years' experience in chemical engineering processes and complex mathematical modeling. He is currently a program manager at Stone & Webster Environmental Services and is responsible for hazards analysis, mixed (hazardous and radioactive) waste, and engineering to minimize hazards. Dr. Doerr has assisted clients with New Jersey's Toxic Catastrophe Prevention Act and California's Risk Management Prevention Program. He has taught hazards analysis and has performed studies for hazards and operability analysis, source term estimation, and risk reduction.

William F. Early, Jr., P.E., is the manager of safety and risk management for Stone & Webster's Process Technologies and Project Services Division in Houston, Texas. He is a licensed professional engineer in the state of Texas. He has an M.S. degree in chemical engineering from the University of Mississippi and has broad process background covering many refining, chemical, and petrochemical designs. His present position includes program management for hazards analyses, risk management programs, and general safety considerations in support of process designs. He also presents training programs to industry in hazards identification techniques, risk management programs, and other related services. Mr. Early is active in the Safety and Health Division of AIChE, serves on an AIChE-CCPS subcommittee, and is a member of SRA and ASSE.

Frank Elia, P.E., holds a B.S. degree in chemical engineering and a master of engineering degree in nuclear engineering, both from Cornell University, and a master of science degree in mechanical engineering from Northeastern University. He has 19 years' experience in plant safety engineering, including chemical process and nuclear power plants, and he is currently supervisor of the Safety Engineering and Analysis Group at Stone & Webster Engineering Corporation in Boston. He is a registered professional engineer in the states of Massachusetts and Nebraska.

Dixie J. Finley has had 8 years' experience in emergency preparedness and is currently responsible for the Hazardous Operations Emergency Response Program at Stone & Webster, which ensures compliance to federal and state regulations. Ms. Finley graduated from the University of California at Davis with an A.B. degree in history and received a B.S. degree in biological conservation from California State University in Sacramento in 1985. Ms. Finley has coordinated drill and exercise programs for nuclear plants, has supervised emergency preparedness training, and has developed OSHA-required training material for the petroleum-refining industry. She is currently participating in the development of a risk management prevention program at a refinery in northern California.

Robert T. Hessian, Jr., P.E., has 12 years' experience in system design and risk assessment and is currently a project manager for environmental projects in the pharmaceuticals, specialty chemicals, and process industries in Stone & Webster's Plant Safety Program. He received a B.E. degree in Mechanical Engineering from Stevens Institute of Technology and has taught and prepared curricula related to risk management program development, hazards analysis, hazards assessment, severe accident analysis, and consequence analysis for Stone & Webster's Plant Safety Program.

Michael C. Livingston is a senior executive consultant with NUS Corporation in Houston, Texas. Mr. Livingston received his B.S. degree in chemical engineering and his M.S. degree in environmental engineering from the University of Arkansas. For more than 17 years, Mr. Livingston has provided safety, process, mechanical, and environmental engineering to consulting, engineering design, and operating companies. He has reviewed and performed hazard and operability (HAZOP) studies, participated in site assessment and hazard identification surveys, consequence analysis, and mitigation programs, and has been active in the development of risk management and prevention programs. Mr. Livingston has also performed conceptual and detailed design engineering of chemical process facilities and petroleum processing facilities for both onshore and offshore installations. Mr. Livingston is a registered environmental professional in Texas, is a registered professional engineer in several states, and is a member of ASSE, WSO, and SRA.

Robert W. Myers is a registered environmental assessor with more than 16 years' experience in the development of multidisciplinary academic, government, and industrial projects. He is a project manager and project engineer for environmental projects in the industrial, water resources, and plant safety business sectors. Mr. Myers has many years of industrial training experience, as both an instructor and a curriculum developer, and he has lectured in emergency planning and hazard identification for Stone & Webster's Chemical Safety Program.

Michael J. Natale is a manager of client services for Stone & Webster Management Consultants, Inc., in New York. Mr. Natale coordinates all risk management consulting assignments. He received a bachelor's degree in economics from Rutgers University and an M.B.A. from Adelphi University. He also holds the ARM designation. Mr. Natale served as Assistant Risk Manager for Suffolk County Government, New York, for seven years and also has experience as a property and casualty underwriter.

Robert O'Mara graduated from Yale University with a B.E. in chemical engineering in 1958, and started his career at Pennsalt (now Pennwalt Corp.) in 1961. He has more than 28 years of experience in chemical engineering. Since 1967, he has been employed at Stone & Webster Engineering Corp. in Boston, specializing in system design, fault tree analysis, probabilistic risk assessment, and project engineering and management on a variety of assignments in chemical, nuclear, and fossil fuel generating stations, and industrial units. Mr. O'Mara has been a member of the

American Institute of Chemical Engineers since 1967 and is a registered professional engineer in Massachusetts, Kentucky, Wisconsin, Pennsylvania, and New York.

Kenneth F. Reinschmidt has been involved with the application of computer technology to engineering, design, and construction for more than 20 years. He received B.S., M.S., and Ph.D. degrees in Civil Engineering from the Massachusetts Institute of Technology. He joined Stone & Webster Engineering Corporation in Boston in 1975 and is currently president of Stone & Webster Advanced Systems Development Services. Dr. Reinschmidt is a member of Chi Epsilon, Tau Beta Pi, Sigma Xi, the Institute of Management Sciences, the Operations Research Society of America, the Society of Manufacturing Engineers, and the American Society of Civil Engineers. He is also Senior Vice-President and member of the board of directors of Stone & Webster Engineering Corporation, and a member of the Board of Directors of Applied Information Technologies S.p.A. of Milan, Italy.

Jack N. Rubin, P.E., is a consulting engineer with Stone & Webster and has 34 years' experience in process industries. His primary experience has been in the area of process design. He is knowledgeable in the design of flexible energy efficient plants and has been awarded U.S. patents in three technologies. A fourth is pending.

Barbara B. Salter, P.E., has had 12 years' experience in the chemical process and power industry. She has had extensive experience in projects involving safety and risk analysis, extremely hazardous chemicals, and waste treatment. As project manager in the Stone & Webster Plant Safety Program, she is responsible for hazards analysis and risk assessments. Ms. Salter has taught hazards assessment and performed hazard and operability studies, as well as dispersion and consequence analysis. Ms. Salter received a B.S. degree in Chemical Engineering from Bucknell University, and is a licensed Professional Engineer.

Michael E. Sawyer is a senior engineer in Stone & Webster's Safety & Risk Management Division. He is a safety engineering graduate of Texas A&M University and a certified safety professional. He has over 8 years' experience in system safety analysis in the chemical and aerospace industries, including extensive fault tree analysis and consequence modeling. Mr. Sawyer has conducted research into the safety aspects of explosion and of fire intensity and in atmospheric dispersion modeling of accidental releases of hazardous materials.

Robert M. Sherrod, P.E., Senior Engineer in Stone & Webster's Safety and Risk Management Division in Houston, has more than 28 years of experience in refining and chemical processing industries. This includes more than 6 years of experience in hazard identification and hazard analysis, 8 years of experience in technical service to operating companies, and more than 14 years of experience with engineering and design companies. Mr. Sherrod has been involved with HAZOP training, facilitation, and evaluation, and has been active in plant audits and safety protocol evaluation for Stone & Webster.

Stephen A. Vigeant is a senior principal environmental scientist with Stone & Webster Environmental Services, Inc., in Boston, Massachusetts. He has over 14 years of experience in environmental impact assessment and licensing/permitting activities. Mr. Vigeant holds a bachelor of science degree in meteorology from Lowell Technological Institute and a master of science degree in meteorology from Pennsylvania State University. He has also been designated a certified consulting meteorologist (CCM) by the American Meteorlogical Society. Mr. Vigeant has published or coauthored seven technical papers dealing with atmospheric dispersion of various forms.

Barbara B. Waterhouse has a masters degree in environmental epidemiology and has experience in interpreting federal, state, and local government regulations dealing with hazardous materials and health and safety. As a consultant on the health effects of environmental pollutants with Stone & Webster Engineering Corp., she interprets epidemiologic and toxicologic data for risk assessment of potentially toxic chemicals and determines applicable regulations for safe disposal.

1

Risk Management Programs

Robert W. Myers, Joseph J. Cramer, and
Robert T. Hessian, Jr.

A risk management program is primarily a management tool. An operating chemical process facility requires a management program that ensures a consistent response to identified risks. An effective risk management program provides responsible administrative control, logical application of the best possible technical analyses, and the involvement of all personnel.

Although risk management programs are now primarily voluntary, their concept is gaining credence and support in regulatory circles. Thus, many companies in the chemical process industry have adopted and implemented effective risk management programs. In this introductory chapter we will review the elements of a good risk management program.

A successful risk management plan must be

- Credible
- Organized
- Thorough (addressing the concerns of the public)
- Relevant
- Doable and economical
- Based on existing technology (with flexibility to adapt to later advances)
- Publicized.

The first four features are closely related. The clearer it is that a program is comprehensive and enforced, the more support it will generate from interested parties. Also, a thorough program identifies potential scenarios for accidental releases, the potential consequences of such releases, and the actions planned to alleviate a problem when it occurs. This imposes a burden on companies using proprietary technology in their processing operations, as it requires documentation to be made available to the public for screening. The fifth and sixth features are also critical, requiring that the program be efficient and flexible. Too much rigidity or costly, impractical elements will ultimately cause the program to fail. A good design must give consideration to human factors and the social sciences.

Key elements that should be present in a risk management program include

- Hazard identification
- Consequence analysis
- Control or treatment responses (management)
- Procedures:
    Operating
    Maintenance
    Testing and inspection
    Change control
- Training
- Emergency planning
- Accident investigation
- Audits.

Most of these elements are examined in more depth in other chapters of this book.

HAZARDS IDENTIFICATION

Techniques should be established at a facility to ensure a rigorous, comprehensive review of material handled, equipment used, and operational procedures utilized. The results of this review should be documented and held for future reference and used when any change is made in the process or operation. Typical hazard identification techniques that may be applicable include the use of checklists, WHAT-IF analysis, hazard and operability studies (HAZOP), fault tree analysis (FTA), or failure modes and effects analysis (FMEA).

Each of these techniques has distinct advantages and disadvantages. None of the aforementioned methods are universally applicable or uniquely correct. The selection of a hazard identification technique for any particular process depends on several factors. Obviously, the stage of development of the process is critical. Different methodologies could be used during the conceptual design, construction, or operating phases of a process. Screening and ranking methodologies are most appropriate in the early stages of process development. Typical approaches include preliminary hazard analyses or the use of hazard checklists or indices, such as Dow/Mond. These are strictly experience-based approaches and help to establish relative levels of potential hazards for a process.

More exacting methods, such as FMEA, WHAT-IF analysis, and HA-ZOP, permit a more structured approach to hazard identification as well as encouraging creativity. These techniques, which are most appropriately implemented in the latter stage of design or during operation of a process, depend on more developed information and documents such as piping and instrumentation drawings (P&IDs). Each of these approaches is more likely to uncover unusual or unexpected events that, although low in probability of occurrence, have potentially severe consequences. Their use requires considerable skill, and they are time-consuming and costly to apply. However, the results of applying these techniques indicate priorities that can assist management in making decisions.

Mathematical and probabilistic techniques such as fault trees and event trees can be very forceful tools, but these require even more expertise and resources. These methods are particularly good for reviewing complex processes in which system interdependencies and interactions exist. Where a sufficient data base exists, these techniques can facilitate quantitative estimates of the frequency of occurrence for specific accident scenarios.

The important thing is to ensure that appropriate methods of hazard identification are applied to new processes, modified projects and, on a periodic basis, to operating processes. Furthermore, the choice of methodologies should follow from an evaluation of all relevant factors, such as the development stage, complexity, project location, local culture or habits, experience of plant personnel, expertise of the evaluators, and the consequences of the hazards.

Both the Chemical Manufacturer's Association (CMA) and the American Institute of Chemical Engineers (AIChE) Center for Chemical and Process Safety (CCPS) have prepared extensive guidelines for the selection of appropriate hazard identification techniques. The World Bank has also prepared a manual to assist in performing hazard evaluations internationally.

CONSEQUENCE ANALYSIS

The consequences of undesired events, identified by a hazard evaluation procedure, must also be determined. With the exception of rough screening analyses (discussed in Chapter 2), consequence analysis is site-specific and must consider the type of hazard involved, site location, population density, and prevailing weather patterns. When the consequences of an undesired event are calculated, both health and economic effects should be considered. For explosions, pressure-wave radii should be calculated; for fires, fireball radii and thermal radiation values versus distance should be determined; for toxic releases, airborne concentrations and hydrological concentrations as functions of distance should be determined.

Consequence analysis is an important part of a risk management program, because risk is defined as the product of the probability of occurrence of an event and its consequences. The management of risk associated with an identified hazard requires the best possible understanding of both the probability of occurrence and the projected consequences resulting from a particular hazard. When included with hazard identification and quantification, the analysis is referred to as a "hazard assessment" in many risk management programs. Good, accurate dispersion modeling or simulation in a realistic fashion is very difficult. Atmospheric dispersion is not readily duplicated in a laboratory, and it is even more difficult to reduce to analytical terms. Efforts in this direction will be increased as the magnitude of both risks and mitigation costs are better understood, but uncertainties will always exist.

CONTROL OR TREATMENT (MANAGEMENT)

Management is at the core of any structured risk management program, and it should proceed from a realization that the potential for accidents will always exist, no matter how many automatic safeguards have been or will be installed. Only by recognizing potential accidents can a good program minimize potential consequences and avoid disasters.

Specific means used to control potential releases or their consequences to the environment must be identified. These could include provision for scrubbing systems to neutralize or remove hazardous components, flare systems or incinerators to destroy hazardous compounds, or even secondary containment devices to temporarily hold the hazardous material until it is further processed before release. Numerous guides and texts exist that can aid in the selection of appropriate equipment or process changes. The principles involved are similar to those encountered in the development and operation of processes. The optimum solution may be to modify the process rather than to add hardware. The real importance of singling out this step is to ensure that appropriate action is taken on every identified hazard. After the probability of occurrence and the potential consequences have been considered, a decision can be made that the risks are acceptable and no further action is required.

Procedures must be integrated into the risk management program for administrative control. Appropriate personnel should be charged with authority to ensure that procedural changes are properly administered, and an organization chart should be made that identifies all personnel at the facility who are assigned appropriate responsibilities to carry out the activities of the risk management program. A control program for handling hazardous chemicals should include a properly defined and regulated set of procedures for operation, maintenance, safety, training, auditing, and investigation. With the concept in mind that a program is designed to prevent releases of chemicals, any changes to existing facility and equipment as well as to processes must be carefully assessed, and effective communication between management, engineering and operations must be maintained.

Engineering control must be exercised, especially in the design, equipment specification, and facility construction stages, to ensure that these activities are carried out in accordance with the appropriate industry codes, consensus standards, and government regulatory requirements.

PROCEDURES

The implementing procedures must be consistent with other elements of the plant safety program. Whenever possible, previously existing safety practices should be integrated into the evolving risk management framework. Operating procedures must address issues revealed in the hazard identification step. Maintenance, testing, and inspection procedures must address hardware items determined to be critical for safety. A preventive maintenance system that identifies trends in equipment failures should be in place to provide data that can be used to prevent serious problems before they occur. Proper procedures may be the most economical and, in some cases, the only practical way of managing a particular risk.

The importance of change control in risk management cannot be over-emphasized. No matter how minor a change may appear to be, control must be maintained by documentation and current drawings. Control must also provide for appropriate hazard analysis as part of any change (documentation prepared in the hazard identification step is useful here). To be successful, this cannot be just a paper program; it must be a controlled process and requires management commitment. The types of procedures applicable are

- Operating procedures that embrace normal facility startup, shutdown, sampling, emergency shutdown, and operating staff shift change
- Maintenance procedures that include normal maintenance scheduling, corrective maintenance, preventive maintenance, and work permit management
- Safety procedures that address plant safety practices (under normal and abnormal operating conditions), abnormality mitigating procedures, emergency response, and first-aid treatment

- Inspection and testing procedures that are related primarily to facility equipment during operation, maintenance, and auditing work
- Communication procedures that are of prime importance, in case of an emergency when a systematic channel of communications is necessary.

As facility operation is an ever-changing process, changes in the above procedures are inevitable. Therefore, the procedural changes for a risk management program must be properly governed and approved by persons assigned by management.

TRAINING

A recent study of accidents in petroleum processing and storage facilities has revealed that roughly two thirds of the accidents were caused by human error rather than by hardware failure or design deficiencies. The best-designed administrative controls are worthless if the people designated to implement those controls have not been trained to understand their intent and proper implementation. Training cannot consist of mere "rubber-stamp" courses; training courses must be integrated into a coordinated program in an atmosphere of corporate commitment. Those trained in the social sciences can probably offer the most to improve the efficiency of training programs, and their insights should be sought during the design and implementation of these training programs.

Training is a collective term for all those activities in which facility personnel are educated to be familiar with the operation and safe practices. General orientation training is the basic education required by all newcomers to a plant. A thorough introduction and a plant tour constitute the fundamental elements of general orientation training. More specific training efforts include on-the-job training, which should include classroom lectures and field-specific practices. Both the classroom lectures and the field-specific practices should be provided to plant operators and principal supervisors to ensure that they understand the basic operation techniques and the underlying principles (in simple language) governing the process. This part of the training should include oral and written performance evaluations.

Safety training must be carefully planned and administered by the safety officer of the corporation. Errors are more readily identified by persons who have been properly trained and acquainted with the correct mitigative procedures. This type of training needs to be well defined, plant-specific, and conducted by a qualified trainer. It should consist of classroom lectures, workshop studies, drills, and exercises, and the scope of such training should include basic plant safety, first-aid treatment information for each of the hazardous materials on-site, emergency responses, hazardous work area precautions, and hazard notification procedures. Training programs are described in more detail in Chapter 15.

Risk management program training courses should be conducted for those persons assigned to the program's development and implementation. The concept of risk management program training is to enhance the awareness of these persons and to support their continuous commitment to program implementation.

Refresher training courses are important to the operators, supervisors, officers, and managers who are accustomed to the existing plant and its day-to-day operation. All facility personnel must be informed of changes that may have taken place at the facility in equipment, in systems, in handling restrictions for a new chemical, or in any other areas that may affect safety. To reduce risks associated with hazardous chemical activity, an appropriately designed refresher training course is necessary once a year. Experience has shown that a one-year interval between normal refresher training courses is optimal. Beyond one year, operator skills begin to deteriorate, especially for activities that are performed infrequently.

To enhance the effectiveness of training programs, it is necessary to document the training procedures, evaluation guidelines, and qualifying criteria for each employee. Should in-house training be insufficient (e.g., if a qualified instructor is lacking), outside training assistance should be sought for the benefit of the program and to ensure a continuum of training activities.

EMERGENCY PLANNING

Despite the best efforts of plant designers, equipment suppliers, and operators, zero risk is an unattainable goal. Results of the consequence analysis can be used as a basis for establishing contingency plans. Personnel must be trained for the roles they will assume in the event of emergencies. Channels of communication and coordination with local government and volunteer agencies must be established. The CMA and the Environmental Protection Agency (EPA) are both active in this area and have prepared worthwhile guidelines for the preparation of effective emergency plans. The Emergency Planning and Community Right-to-Know Act of 1986 now

requires nearly all companies to participate in the preparation, maintenance, and potential implementation of comprehensive emergency plans.
Training and practice, with concentration on the proper use of feedback,
should be emphasized.

Emergency preparedness is a preplanned system that specifies how a
well-defined emergency organization will function in emergency situations. The emergency organization must have a clearly identified organization chart listing its members and their responsibilities in the hierarchy.
The emergency organization should be responsible for carrying out the
following tasks:

- Developing an emergency response plan for the facility
- Implementing and reviewing the emergency response plan
- Developing emergency response procedures for implementation of
  the plan
- Training the emergency response organization by means of lectures,
  workshop studies, drills, and exercises
- Establishing and operating an emergency command center
- Coordinating on-site emergency activities
- Maintaining communication with state and local agencies on a 24-hour
  basis
- Controlling inventory and reviewing resources for emergency re-
  sponses, such as emergency lighting, radio equipment, firefighting
  equipment, and first-aid equipment.

Practice of emergency responses is as vital as the setup of the emergency
organization. Timely execution is critical during a real emergency. No
matter how well the response team is organized and equipped, the team
members will not perform as intended unless they have had practical
experience under simulated emergency conditions. The only sensible way
to get experience is through on-site training, which has been designed
specifically for the facility and conducted by well-qualified, experienced
instructors. Emergency response training is different from other safety
training in which most of the responses and actions are conclusions from
previous experience. Theories and principles are useful only when team
members have had sufficient time to think about the appropriate actions
and responses in advance of actual events. Effective emergency response
training includes classroom training on the plan and procedures to be
followed by practical hands-on practice under preplanned, simulated,
emergency conditions. This hands-on training is usually in the form of

drills or exercises in which the response organization has the opportunity to implement all or a portion of the plans and procedures under controlled conditions. Areas to be covered under this type of training include

- Initial response and mitigation
- Notification and activation of the organization
- Setup and operation of command center
- Communication systems
- Coordination with federal, state, and local responders.

Lessons learned during this training should be documented and used in revising the plan, procedures, and training program to ensure maintenance of an ongoing, effective emergency response capability.

ACCIDENT INVESTIGATION

For experience to be an effective teacher, accurate and unbiased feedback must be obtained from the experience. Unfortunately, an accident tends to make the people involved very defensive and fearful. An effective accident investigation procedure must be perceived as a learning tool, to determine possible preventive measures, rather than as a search for a scapegoat. Human nature being what it is, this is easy to accept in theory, but often very difficult to achieve in practice. Nevertheless, for accident or near-accident investigation to work effectively, management must practice what it preaches. The best programs are those in which people can learn from their mistakes. Workers must be encouraged to report their mistakes, and should not be disciplined unless the offenses are repeated.

The objectives of accident investigation are

- To determine the cause and effect,
- To provide documentation for personnel education to prevent recurrence under similar circumstances, and
- To record the measures taken for remedial action.

When conducting an accident investigation, a responsible person or, in the case of a serious accident, a team consisting of management and supervisors of different interests should be established as soon as possible, so that reliable evidence and data can be collected to facilitate the investigation process. Information can be gathered from the scene, persons responsible can be interviewed, technical data can be inspected, and historical data on similar accidents can be studied.

Proper investigation procedures shall be laid down for the risk management program so that qualified persons can conduct a comprehensive investigation. All information associated with the accident should be properly documented and maintained. An accident investigation report should provide a historical record of the facility's operation and adequately document the investigation, the conclusions, recommended actions, and appropriate resolutions. Reports should not be closed until all remedial actions have been completed.

AUDITS

Periodic checks are required to determine the effectiveness of the program as well as the degree of compliance. Management that pays lip service to safety and related programs, yet concentrates its resources on short-term production goals, is not doing itself or the industry any favors. Audits are a key resource for feedback to management on program effectiveness and adequacy, as well as a signal to corporate personnel and the public that the risk management program has continued corporate commitment.

Without a proper audit program, management cannot properly implement a risk management program. Risk management program auditing is the chief administrative control through which implementation and enforcement are ensured.

Audit programs can be administered in any number of ways. Programs can be administered from central audit departments or run by ad hoc committees, which can use personnel from various plants or groups. Outside consultants can be used entirely or in part, or some combination of personnel can be used.

Audit program requirements are specific to each program as well as to a plant's characteristics. The scope of an audit program may range from a review of normal operation and maintenance procedures to a detailed inventory of records and checklists, depending upon the complexity of the risk management program. However, the audit program should ensure the review of each element for its applicability and implementation and should include recommendations for remedial action.

The audit process can take different forms. These include, but are not limited to, physical inspections, documentation and records review, administrative controls for tracking and checking program implementation, as well as interviews with appropriate personnel. The frequency of auditing depends on the program and facility, but a yearly audit is recommended for typical facilities.

CURRENT AND PENDING REGULATION AND
INDUSTRY GUIDANCE

State Regulations

Legislators, regulators, and industry professionals are becoming increasingly active in developing risk management program descriptions, requirements, and guidance. New Jersey, California, and Delaware already have statutory or regulatory requirements, or both, specific to risk management of chemical processes. The reader is referred to the respective statutes and regulations for a detailed discussion of requirements for each jurisdiction. Illinois, Texas, and other states are exploring similar regulatory or statutory approaches to the management and control of process hazards.

Federal Clean Air Act Amendments (CAAA)

The recently adopted CAAA of 1990 includes the following requirements for risk management programs:

- Chemical Safety Hazard Investigation Board
- Extremely hazardous substances defined (initial mandatory list)
- Current state statutes not to be superseded
- Hazard screening
- Consequence analysis
- Hazard mitigation
- Program compliance authority for EPA (civil penalties).

Occupational Safety and Health Administration (OSHA)

The Federal Occupational Safety and Health Administration has issued proposed regulations for "Process safety management of highly hazardous chemicals." The proposed regulations, 29 CFR 1910.119, address the following topic areas:

- Process safety information
- Process hazard identification

- Operation procedures
- Training
- Contractors
- Pre-startup safety review
- Mechanical integrity
- Hot work permits (maintenance work on operational or energized systems)
- Management of change
- Incident investigations
- Emergency response.

Industry Guidelines

The American Petroleum Institute (API) has recently published their recommended practice guidelines (API-750), entitled "Management of Process Hazards." This guidance documentation applies to refineries, petrochemical operations, and major processing facilities as a consensus standard. The objective of the standard is the prevention of catastrophic releases through the development of management control systems that address the following areas:

- Process safety information
- Process hazards analysis
- Management of change
- Operating procedures
- Safe work practices
- Training
- Assurance of the quality of mechanical integrity of critical equipment
- Pre-startup safety review
- Emergency response and control
- Investigation of process related incidents
- Audits of process hazards management systems.

SUMMARY

All companies can adopt, implement, and eventually institutionalize an effective corporate risk management program. No one should realistically expect any long-term resolution of either the perceived or real problem of chemical risks without a major initiative from within the industry. The industry is utilizing analysis techniques developed in other industries, utilizing the expertise of those social sciences accustomed to dealing with human factors, and establishing organizations modeled on those used successfully in other fields. These mechanisms are being adapted to fit the chemical industry.

To proceed with these initiatives, the industry has achieved a real and substantial consensus. The costs incurred through application of a comprehensive risk management program must be accepted as a necessary cost of doing business and serve as an incentive to design processes and plants which are inherently safer. It is entirely possible that the careful adherence to safety principles may, in the long-term, reduce wastes and improve productivity.

SUGGESTED READINGS

American Petroleum Institute. 1990. *Management of Process Hazards API Recommended Practice 750,* First ed. Washington, DC: API.

Center for Chemical Process Safety. 1985. *Guidelines for Hazard Evaluation Procedures.* New York: American Institute of Chemical Engineers.

European Council of Chemical Manufacturers' Federations. 1986. CEFIC views on the quantitative assessment of risks from installations in the chemical industry. Brussels, Belgium.

Garrison, W. G. 1988. Major fires and explosions analyzed for 30-year period. *Hydrocarbon Processing,* 67(9): pp. 115–120.

Haasl, D. F., et al. 1981. *Fault Tree Handbook,* Washington, DC: USNRC, NUREG-0492.

Institution of Chemical Engineers. 1985. Risk Analysis in the Process Industries. Rugby, England: European Federation of Chemical Engineering Publication No. 45.

International Confederation of Free Trade Unions and the International Federation of Chemical, Energy and General Workers' Unions. 1985. The trade union report on Bhopal. Geneva, Switzerland: ICFTU-ICEF.

Knowlton, R. E. 1981. Hazard and operability studies, the guide word approach. Vancouver, BC: Chemetics International Company.

Occupational Safety and Health Administration. 1988. Systems safety evaluation of operations with catastrophic potential. Washington, DC: OSHA Instruction CPL 2-2.45.

Organization Resources Counselors, Inc. 1988. Process hazards management of substances with catastrophic potential. Washington, DC: Process Hazard Management Task Force.

Process Safety Management. 1985. Washington, DC: Chemical Manufacturers Association.

Prugh, R. W. 1987. Evaluation of unconfined vapor cloud explosion hazards. International Conference on Vapor Cloud Modeling. J. L. Woodward, Ed.

Center for Chemical Process Safety of the American Institute of Chemical Engineers/Institution of Chemical Engineers/U.S. Environmental Protection Agency, Cambridge, MA, pp. 712–755.

State of California, Office of Emergency Services, 1989. Guidance for the Preparation of a Risk Management and Prevention Program. Sacramento, CA.

State of Delaware, Department of Natural Resources and Environmental Control. 1989. Regulation for the management of extremely hazardous substances and background and information document, extremely hazardous substances risk management act. Dover, DE.

World Bank. 1985. *Manual of Industrial Hazard Assessment Techniques*. London: World Bank.

2

Screening Analysis Techniques

Michael E. Sawyer, Michael C. Livingston, and William F. Early

Over the years the insurance industry, along with the chemical process industries, has developed various hazard indices. These indices generally serve as initial risk assessment tools to identify hazardous operations and processes. Often these hazard indices have been developed in direct response to catastrophic events.

As a result of catastrophic occurrences over the last few years, the chemical process industry has experienced a heightened awareness of the consequences arising from chemical process plant accidents. People living or working around these plants are demanding greater protection for themselves as well as for their environment.

Several hazard indices and screening analysis techniques have been used by industry to determine the risk associated with a unit operation, process, or plant, thereby allowing a ranking or comparative analysis of those risks. Many of the indices are interrelated and often complement one another during a risk assessment project. The following sections address some of the more prevalent ones used in the process industry.

DOW FIRE AND EXPLOSION INDEX

The need for a system to identify chemical process areas with significant loss potential led Dow Chemical Company to develop a Fire and Explosion Index[1]. The first edition was released in 1964 and has undergone several revisions since then. The concept of this index is to separate the system or plant into manageable sections in preparation for analysis. The key to its usage is to identify the material(s) in the section being analyzed, select the dominant chemical parameters, and assess their thermodynamic properties.

The purpose of the index is to define a methodology that can

- Quantify anticipated damage from potential fires and explosions in realistic terms
- Identify contributors to potential fire or explosion incidents
- Communicate the risk potential to management in terms of dollars.

The underlying objective is that, upon completion of this index, the analyst should be aware of the realistic loss potential of the process under consideration. However, the index does not purport to be as comprehensive as a detailed hazard analysis and is therefore included as a screening technique.

To conduct an assessment using the index, the analyst must have, as a minimum

- An accurate layout of the plant or unit under consideration
- Process flow diagrams
- Cost data for the plant or unit under consideration
- Appropriate forms and summary sheets as identified in the sixth edition of the *Dow Fire and Explosion Hazard Classification Guide*.

The basic procedure for index calculation consists of the following steps:

1. Select pertinent process.
2. Determine unit hazard factor:
     Calculate general process hazard factors.
     Calculate special process hazard factors.
3. Calculate fire and explosion index:
     Determine exposure radius.
4. Conduct risk analysis:
     Calculate damage factor.
     Determine value of equipment in exposure area.
5. Determine base maximum probable property damage (MPPD).
6. Calculate loss control credit factors.
7. Calculate actual MPPD.
8. Determine maximum probable days outage and business interruption.

A typical polyol reactor batch process is used to illustrate how the index is applied. The example batch process uses an initial charge of glycerine to the reactor and then progressively adds propylene oxide and ethylene oxide to form the polyol.

After the process for study is defined, the material factor is determined, a predictive value of the energy potential of the chemicals contained within a process unit. The factor is selected from a matrix developed by using the reactivity and the flammability numbers of the materials and compounds handled within the process unit. The materials can be liquid, gas, or solid, and either a pure compound or a mixture. The suggested material factor for this example is propylene oxide (PO) because of the residual PO present during the batch process.

After the appropriate material factors have been determined, the unit hazard factor is calculated. This numerical value is determined after the penalties assigned to the "general process hazards" and "special process hazards" have been calculated.

The general process hazards are factors based upon the type of reaction occurring within the process unit. The more violent the reaction, the larger the penalty (that is, the larger the factor). An exothermic reaction has a higher penalty than an endothermic reaction. Nitration reactions have the largest exothermic penalty, while hydrogenation has the minimum penalty. Calcination and pyrolysis usually have equal endothermic reaction penalties. The example process was assigned the maximum penalty of 1.25 because of the potential for a violent reaction.

Also included is a component factor derived from material handling and transfer operations, enclosure of process units, easy access and egress to the units or equipment, and type of surface drainage. The summation of the general process hazards for the polyol example is 3.55, as shown in Figure 2-1.

The special process hazards are factors that contribute primarily to the probability of an incident. The subsections of this factor are based on the physical and mechanical parameters associated with the process. Some of the parameters are the proximity of the process temperature to the flash point or boiling point; operating near the flammability range with equipment that can bring the product into contact with air (i.e., barge loading with no purging and no inert gas padding). Other considerations include the set pressures of safety valves; use of fired heaters, rotating equipment, or hot oil systems; and quantity of materials stored or contained within the processing area. The total of these factors for the polyol example equates to 4.14.

The unit hazard factor is the product of the general process hazards and the special process hazards. This factor usually has a range of 1 to 8. Even if a higher value is obtained, a maximum of 8 is applied to this factor. The product of the polyol example is greater than the maximum; thus 8 was used as the unit hazard factor.

The fire and explosion index is the product of the unit hazard factor and the material factor. This value is used for estimating the damage that would

FIRE AND EXPLOSION INDEX

LOCATION _____ DATE _____

PLANT _____ PROCESS UNIT _____ EVALUATED BY _____ RECEIVED BY _____

MATERIALS AND PROCESS

MATERIALS IN PROCESS UNIT_____

STATE OF OPERATION _____ BASIC MATERIAL(S) FOR MATERIAL FACTOR

MATERIAL FACTOR (SEE TABLE I OR APPENDICES A OR B)		24

	Penalty	Penalty Used
1. GENERAL PROCESS HAZARDS		
BASE FACTOR	1.00	1.00
A. EXOTHERMIC CHEMICAL REACTIONS (FACTOR .30 TO 1.25)	1.25	
B. ENDOTHERMIC PROCESSES (FACTOR .20 TO .40)		
C. MATERIAL HANDLING AND TRANSFER (FACTOR .25 TO 1.25)	0.85	
D. ENCLOSED OR INDOOR PROCESS UNITS (FACTOR .25 TO .90)	0.45	
E. ACCESS	0.35	
F. DRAINAGE AND SPILL CONTROL (FACTOR .25 TO .50) 200 Gals.		
GENERAL PROCESS HAZARDS FACTOR (F_1)	3.55	
2. SPECIFIC PROCESS HAZARDS		
BASE FACTOR	1.00	1.00
A. TOXIC MATERIAL(S) (FACTOR 0.20 TO 0.80)	0.40	
B. SUB -ATMOSPHERIC PRESSURE (< 500mm Hg)	0.50	
C. OPERATION IN OR NEAR FLAMMABLE RANGE INERTED/NOT INERTED		
1. TANK GARMS STORAGE FLAMMABLE LIQUIDS	0.50	
2. PROCESS UPSET OR PURGE FAILURE	0.30	
3. ALWAYS IN FLAMMABLE RANGE	0.80	0.80
D. DUST EXPLOSION (FACTOR .25 TO 2.00) (SEE TABLE II)		
E. PRESSURE OPERATING PRESSURE 100psig		
RELIEF SETTING 125psig	0.34	
F. LOW TEMPERATURE (FACTOR .20 TO .30)		
G. QUANTITY OF FLAMMABLE/UNSTABLE MATERIAL: QUANTITY 100 lbs. H_c = 13.2 BTU/lb.		
1. LIQUIDS, GASES AND REACTIVE MATERIALS IN PROCESS (SEE FIG. 3)		1.60
2. LIQUIDS OR GASES IN STORAGE (SEE FIG. 4)		
3. COMBUSTIBLE SOLIDS IN STORAGE. DUST IN PROCESS (SEE FIG. 3)		
H. CORROSION OR EROSION (FACTOR .10 TO .75)		
I. LEAKAGE-JOINTS AND PACKING (FACTOR .10 TO 1.50)		
J. USE OF FIRED HEATERS (SEE FIG. 6)		
K. HOT OIL HEAT EXCHANGE SYSTEM (FACTOR .15 TO 1.15) (SEE TABLE III)		
L. ROTATING EQUIPMENT	0.50	
SPECIAL PROCESS HAZARDS FACTOR (F_2)	4.14	
UNIT HAZARD FACTOR (F_1 X F_2 = F_3)	8.00	
FIRE AND EXPLOSION INDEX (F_3 X MF = F& EI)		192

Figure 2-1   Fire and explosion index.

probably result from an incident involving the process under study. The index calculated for the polyol example was 192.

Next, a unit analysis of the process is conducted. The calculated index is converted to a "radius" by multiplying it by 0.84. This radius represents the area containing equipment that could be exposed to a fire or explosion generated by the process under study. The polyol example resulted in an "area of exposure" radius of 161 feet.

The unit damage factor is correlated from the unit hazard factor and the material factor. This factor represents the overall effect of fire and over-pressure damage resulting from an incident. This factor is more of a probability or percentage factor than an actual damage indicator. The polyol example resulted in a damage factor of 0.88.

The credit factor, used in the unit analysis summary, is derived by assigning credits for equipment and procedures installed to prevent or mitigate the effects of a fire or explosion. This factor is a product of three sets of safety-related systems: level of process control, degree of materials isolation, and types of fire protection, and is shown in Figure 2-2. The unit analysis summary for the polyol example is shown in Figure 2-3.

The total production outage is a probability factor derived from the loss of the production capacity of the unit or plant expressed in days. Loss of production is expressed as equipment loss (property damage) and the number of days of production outage.

MOND FIRE, EXPLOSION, AND TOXICITY INDEX

The Mond Division of the Imperial Chemical Industries, Ltd. (ICI) acknowledged that the Dow Fire and Explosion Index could be used as a useful analytical tool for the evaluation of chemical processes. However, ICI revised the index to include toxicity considerations and "offsetting features," which was an expansion of the Dow credit factors. An important outcome of using the Mond Index is that the technique raises questions concerning hazard potential at an early stage in planning before the process is in operation.

The Mond Index is most useful during the planning or engineering phase of a chemical or processing plant. Modifications can be made as needed to reduce the Mond Index value at the least possible cost. During the early stages of the project, the Mond Index would recommend the following objectives for reducing the plant risk assessment:

1. Equipment with the greatest potential hazard should be located away from operating personnel and surrounding civilians.

LOSS CONTROL CREDIT FACTORS

1. PROCESS CONTROL (C_1)

a)	EMERGENCY POWER	.98	f) INERT GAS	.94 to .96
b)	COOLING	.97 to .99	g) OPERATING INSTRUCTIONS/	
c)	EXPLOSION CONTROL	.84 to .98	PROCEDURES	.91 to .99
d)	EMERGENCY SHUTDOWN	.96 to .99	h) REACTIVE CHEMICAL	
e)	COMPUTER CONTROL	.93 to .99	REVIEW	.91 to .98

C_1 TOTAL   0.84

2. MATERIAL ISOLATION (C_2)

a)	REMOTE CONTROL VALVES	.96 to .98	c) DRAINAGE	.91 to .97
b)	DUMP/BLOWDOWN	.96 to .98	d) INTERLOCK	.98

C_2 TOTAL   0.91

3. FIRE PROTECTION (C_3)

a)	LEAK DETECTION	.94 to .98	f) SPRINKLER SYSTEMS	.74 to .97
b)	STRUCTURAL STEEL	.95 to .98	g) WATER CURTAINS	.97 to .98
c)	BURIED TANKS	.84 to .91	h) FOAM	.92 to .97
d)	WATER SUPPLY	.94 to .97	i) HAND EXTINGUISHERS/	
e)	SPECIAL SYSTEMS	.91	MONITORS	.95 to .98
			j) CABLE PROTECTION	.94 to .98

C_3 TOTAL   0.74

CREDIT FACTOR = C_1 X C_2 X C_3 =   0.56

Figure 2-2   Loss control credit factors.

Locate high-risk units near the center of the facility and separate the higher risk units with units of lesser risk as a buffer. The units with the highest potential of aerial explosion from vapor cloud detonation should not be located near property boundaries. Buildings housing personnel should be located near the perimeter of the plant site. Where possible, high-occupancy buildings should be shielded from high-risk units with low-risk units or with low-occupancy buildings.

UNIT ANALYSIS SUMMARY		
FIRE & EXPLOSION INDEX:	192	
RADIUS OF EXPOSURE:	161	
VALUE OF AREA OF EXPOSURE:		$MM: 15.0
DAMAGE FACTOR:	0.88	
BASE MMPD:		$MM: 13.2
CREDIT FACTOR:	0.56	
ACTUAL MMPD:		$MM: 7.4
DAYS OUTAGE:		28/110 DAYS
BUSINESS INTERRUPTION:		$MM: 2.8

Figure 2-3   Unit analysis summary.

2. Space units so as to minimize the effects of an explosion or toxic release from one unit to an adjacent unit.

Pipeways connecting units within the plant should be protected to prevent fire from migrating between units. These pipeways should be protected from vehicular movements and should be located to minimize the impact from a fire or explosion from adjacent, higher elevation, high-risk units. Pipeways should be installed only where absolutely needed.

Units should be isolated away from ignition sources, such as furnaces, flares, and switchgears.

3. Arrange a plant to minimize the effects of an explosion or toxic release on adjacent property.

Storage facilities should be located remote from the process equipment and away from highways, rail traffic, civilian housing, and public access facilities where possible.

4. Locate equipment so that similar, small-potential hazards are located together, or so that a large-potential hazard is broken into smaller units to minimize the effects of an explosion or toxic release.

The units should be separated so that the thermal effects of a fire or the pressure wave and projectiles from an explosion will have minimal impact on the adjacent units.

5. Provide unrestricted access for rescue and firefighting personnel.

Two access/egress corridors for the plant site and for each individual unit within the plant should be available. Access corridors throughout the plant should be wide enough for all sizes of emergency and fire-protection equipment.

Applying a Mond Index calculation to an existing plant would require additional safety modifications or equipment retrofitting to achieve a lower Mond number. To perform a Mond Index, the following steps would be required:

1. Break the plant down into discrete units or processes.
2. For each unit, identify the hazards of the materials entering or exiting the unit, the equipment, and the process.
3. Rate each component of the unit identified.
4. Review the acceptability of the identified hazards.
5. Mitigate the identified hazards to lower the degree of hazard or explore the accuracy of the information used in the development of the unit rankings.
6. Prepare a listing of credits for reducing the overall risk factor (these are design considerations such as level of fire protection, spacing between high-risk pieces of equipment, spacing between control buildings and the equipment, level of instrumentation, fire walls, water curtains, and steam curtains).

The selection of the level of risk can be subjective. If a risk level is too high, a reduction credit can be included to lower the risk level to a more acceptable value.

The Mond Index is a two-part calculation. The initial calculation represents the plant's worst-case hazard rating, while the second calculation represents the corrected hazard rating. Initially, the plant is evaluated with no deduction taken for plant fire protection and safety equipment. The second calculation includes all of the plant protective equipment. The difference in the Mond Index values implies a level of risk reduction that is attributed to the installed fire and safety equipment.

If the second Mond Index value is perceived to be unacceptably high, the basis of the index should be reevaluated. Additional screening techniques and identification methodologies can identify hazard potentials, and with physical plant or operating modifications, the factors used in the calculation of the Mond Index could be lowered to a more acceptable level.

GENERAL SCREENING ANALYSIS TECHNIQUES

The degree of hazard potential associated with a chemical plant or unit can be considered a function of the materials and the operation or process of those materials. An initial screening of each materials hazard potential or their contribution to the overall risk of the plant or unit should be conducted for each process or operation of the plant. Thus, by initially screening out relatively unimportant chemicals and processes, the amount of detailed hazard analysis effort can be reduced significantly without sacrificing confidence in the analysis. The purpose of a chemical hazard screening is to identify the materials or operations that constitute potential hazards to personnel or chemical processes.

The importance measure of a chemical's hazard potential can be determined by its contribution to risk. Since risk cannot be completely assessed until the first iteration of analysis is completed, a screening technique is employed as a preliminary assessment tool. The primary measures used in the screening are the materials' physical and chemical properties, such as flammability, reactivity, and toxicity. This information can be obtained from several sources of varying complexity, depending upon the degree of screening desired.

Several indices, reference sources, and data bases are available to assist in the organization of the materials in accordance with their associated hazards. The following is a partial listing of suggested sources that have developed indices, data bases, and hazard rating criteria for chemical substances:

Substance Hazard Index (API RP 750, Management of Process Hazards)

System of Hazard Identification; Hazardous Chemicals Data (National Fire Protection Association NFC 49)

American Conference of Governmental Industrial Hygienists Threshold Limit Values and Biological Exposure Indices

Extremely Hazardous Substances List and Threshold Planning Quantities (Environmental Protection Agency; 40 CFR Parts 355)

CAMEO Response Information Data Sheets (National Oceanic and Atmospheric Administration).

As with any methodological analysis technique, several steps should be followed when conducting an initial screening. However, before the screening analysis can begin, the analyst should establish some basic parameters regarding what constitutes a hazardous material.

The analyst may opt to identify a data base, such as API RP 750 or 40 CFR 355, if the objective of the screening is to consider any chemical in inventory hazardous if it appears on a predetermined list. The appropriate federal, state, and local regulations should always be consulted before the final selection of a data base or source is made.

The parameters that are used in determining whether a chemical substance is hazardous can vary depending upon the process, operation, proximity, or frequency of the operation. In general, the most significant parameters to be considered are as follows:

1. *Toxicity.* Hazardous chemicals that readily vaporize upon contact with the atmosphere, those that are lethal at low concentrations, carcinogens, and those that are easily ingested are the most hazardous.
2. *Flammability.* Pyrophoric materials, liquids with low flash points and ignition points, and substances capable of exothermic reactions should be ranked relative to their hazard potential.
3. *Reactivity.* Strong oxidizers, organics, and materials capable of detonation or deflagration are considered in this category.

The next step is the chemical inventory identification process. A comprehensive listing of all chemicals applicable to the operation, process, or peculiar system is prepared. This listing must also consider the raw materials, by-products, intermediates, and final products.

Next, the properties of each chemical identified in inventory are compared with the information provided by the selected index, data base, or reference source. After all information has been gathered, the list of chemicals should be consolidated to ensure complete incorporation into the hazard analysis effort. A concise flow diagram of a typical chemical screening process is represented in Figure 2-4.

DATA BASES AND HAZARD CRITERIA

The following sources represent predetermined data bases and hazard evaluation criteria that can assist in the screening process. A brief summary for each identified source is included to assist the analyst in selecting which source may serve the screening application best. However, the following list is only a partial list of suggested sources to assist in screening.

SUBSTANCE HAZARD INDEX

The substance hazard index (SHI) distinguishes chemical substances by their vapor pressures and toxicity ratings[2]. It postulates that a high-vapor-

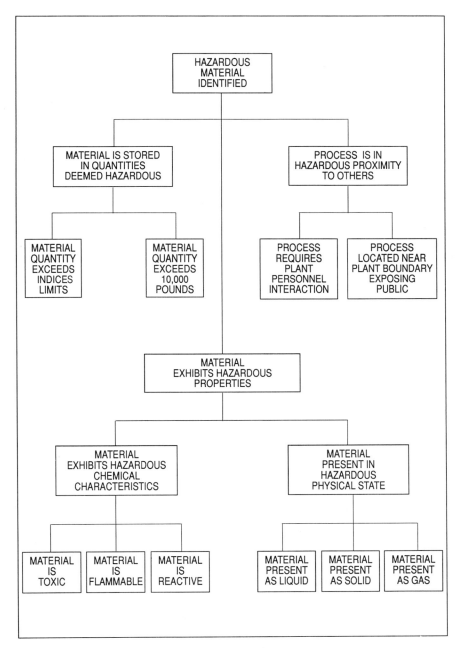

Figure 2-4 Flow diagram.

pressure substance disperses more readily into the atmosphere, and a highly toxic substance release can pose danger at low emission rates. The following expression is used to determine the SHI for chemicals:

$$SHI = EVC/ATC$$

EVC (equilibrium vapor concentration) is the substance vapor pressure at 20 C, in millimeters of Hg, multiplied by 106, then divided by 760. ATC (acute toxicity concentration), in parts per million, is defined as the lowest reported concentration, based on recognized data, that will cause death or permanent injury to human beings after a single exposure of 1 hour or less.

API Recommended Practice 750, Management of Process Hazards, applies to substances that have an SHI greater than 5,000 and are present in quantities above a threshold amount. Table 2-1 is an illustrative list of substances and corresponding SHIs.

OSHA has expressed its intent to use the American Industrial Hygiene Association Emergency Response Planning Guidelines (ERPG-3 level) rather than the API's ATC in pending regulations 29 CFR Part 1910.119 Process Safety Management of Highly Hazardous Chemicals. This approach is very similar to the API approach, but will probably result in a few additional chemicals being listed above the SHI 5,000 plateau.

EXTRAORDINARILY HAZARDOUS SUBSTANCE

The New Jersey Toxic Catastrophe Prevention Act (TCPA) identified a list of chemical substances that if released into the environment would produce a significant likelihood that persons exposed would suffer acute health effects resulting in death or permanent disability[3]. This partial list of substances is known as the extraordinarily hazardous substance (EHS) list, as shown in Table 2-2.

Table 2-1   Substance Description

CAS Number	Substance	Acute Toxicity Concentration	Substance Hazard Index
107-02-8	Acrolein	3.00	97807
814-68-6	Acrylyl chloride	2.40	164474
107-05-1	Allyl chloride	29.00	13793
107-11-9	Allylamine	13.80	18402
7664-41-7	Anhydrous ammonia	1,000.00	8447
7784-42-1	Arsine	6.00	2500000

Table 2-2   Extraordinarily Hazardous Substance

Name of EHS	CAS No.	Reportable Minimum Quantity in Pounds
Hydrogen Chloride	7647-01-0	2,000
Hydrochloric Acid 36% by weight HCl	7647-01-0	5,600
Allyl Chloride	107-05-1	2,000
Hydrogen Cyanide	74-90-8	500
Hydrogen Fluoride (HF)	7664-84-1	500
Hydrofluoric Acid 70% by weight or more HF	7644[84-1] 39-3	700
Chlorine	7782-50-5	500
Phosphorous trichloride	7719-12-2	500
Hydrogen Sulfide	7783-06-4	500
Phosgene	75-44-5	100
Bromine	7726-95-6	100
Methyl isocyanate	624-83-9	100
Toluene-2 4-diisocyanate	584-84-9	100

EXTREMELY HAZARDOUS SUBSTANCE

The Environmental Protection Agency (EPA) publishes a list of extremely hazardous substances in relation to the Emergency Planning and Community Right-to-Know Act[4]. This list was established by the EPA for the identification of chemical substances that could cause serious irreversible health effects from accidental releases. The criteria used to identify the following substances mainly consider acute toxicity, physical form, and how the chemical is used. Table 2-3 illustrates a section of the EHS list.

NFPA SYSTEM OF HAZARD IDENTIFICATION

The National Fire Protection Association (NFPA) has developed a hazard identification system, Number 704, designed to provide the necessary life safety information for emergencies involving chemicals[5]. This system is designed to provide emergency response personnel with general information about chemical hazards during inadvertent releases, fires, and spills. While not specifically designed as a screening tool, this system can provide the analyst with useful information regarding the health, flammability, and reactivity effects of various chemicals. Each chemical listed in this system is identified by a subjective severity ranking of 0 to 4 for health, flammabil-

Table 2-3   Extremely Hazardous Substances

CAS No.	Chemical Name	Reportable Quantity (lb)	Threshold Planning Quantity (lb)
75-86-5	Acetone Cyanohydrin	10	1,000
1752-30-3	Acetone Thiosemicarbazide	1	1,000/10,000
107-02-8	Acrolein	1	500
79-06-1	Acrylamide	5,000	1,000/10,000
107-13-1	Acrylonitrile	100	10,000
814-68-6	Acrylyl Chloride	1	100
111-69-3	Adiponitrile	1	1,000
116-06-3	Aldicarb	1	100/10,000
309-00-2	Aldrin	1	500/10,000
107-18-6	Allyl Alcohol	100	1,000
107-11-9	Allylamine	1	500
20859-73-8	Aluminum Phosphide	100	500
54-62-6	Aminopterin	1	500/10,000
78-53-5	Amiton	1	500
3734-97-2	Amiton Oxalate	1	100/10,000
7664-41-7	Ammonia	100	500
16919-58-7	Ammonium Chloroplatinate	1	10,000
300-62-9	Amphetamine	1	1,000
62-53-3	Aniline	5,000	1,000
88-05-1	Aniline, 2,4,6-Trimethyl	1	500
7783-70-2	Antimony Pentafluoride	1	500
1397-94-0	Antimycin A	1	1,000/10,000
86-88-4	ANTU	100	500/10,000
1303-28-2	Arsenic Pentoxide	5,000	100/10,000
1327-53-3	Arsenous Oxide	5,000	100/10,000
7784-34-1	Arsenous Trichloride	5,000	500
7784-42-1	Arsine	1	100
2642-71-9	Azinphos-Ethyl	1	100/10,000

ity, and reactivity, with 4 representing the most hazardous. Following is an example of this ranking and an interpretation.

ACETIC ANHYDRIDE < 2,2,1 >

Health (2): Materials hazardous to health, but areas may be entered freely with self-contained breathing apparatus.

Flammability (2): Liquids that must be moderately heated before ignition will occur and solids that readily give off flammable vapors.

Water spray may be used to extinguish the fire because the material can be cooled to below its flash point.

Reactivity (1): Material which in themselves are normally stable but which may become unstable at elevated temperatures and pressures or which may react with water with some release of energy but not violently. Caution must be used in approaching the fire and applying water.

Special Information: With reactivity of 1; In addition the noted hazards, the material may react vigorously but not violently with water.

Only a select few of the available indices and screening techniques have been discussed. However, the examples chosen were selected to provide the analyst with a general understanding of how indices and screening techniques can be successfully used as initial assessment tools. As previously noted, there are several screening techniques and indices from which the analyst can choose. Each has unique features and should be chosen with regard to the objectives of the analysis, type of process under study, time, and resources available. The careful selection and combination of techniques discussed in this chapter will enhance the analysis process and provide the necessary credibility.

REFERENCES

1. Dow Chemical Company. 1987. *Fire and Explosion Index Hazard Classification Guide*. Sixth ed. New York: AICHE.
2. American Petroleum Institute. 1990. *Management of Process Hazards*, Recommended Practice 750. First ed., Washington, DC: API.
3. New Jersey Environmental Protection Agency. 1987. Toxic Catastrophe Prevention Act. Princeton, NJ
4. Environmental Protection Agency. 1987. 40 CFR Part 355, "Extremely Hazardous Substances List."
5. National Fire Protection Association 704. *Identification of the Fire Hazards of Materials*. 1989. Quincy, Mass.: NFPA.

3

Checklist Reviews

Robert T. Hessian, Jr. and Jack N. Rubin

A checklist is a list of questions about plant organization, operation, maintenance, and other areas of concern. Historically, the main purpose for creating checklists has been to improve human reliability and performance during various stages of a project or to ensure compliance with various regulations and engineering standards. Each item can be physically examined or verified while the appropriate status is noted on the checklist. Checklists represent the simplest method used for hazard identification.

DESCRIPTION OF METHODOLOGY

As a hazard identification technique, checklists are used as a means of verifying that various requirements have not been neglected or overlooked. Therefore, an audit checklist is usually used after the subject activity (e.g., conceptual design, design development, construction) has been completed. Checklists are based primarily upon the preparer's prior experience, but they can also be based on codes and standards.

When generating an effective checklist, the preparer must recognize additional areas of expertise to be included. For example, a checklist that verifies design completion would include electrical, controls, mechanical, and civil-structural aspects, as well as process-specific attributes such as explosive limits, reactivity threshold, or flammability. Once all the areas of expertise have been identified, arrangements should be made to secure the services of competent personnel in each of those areas. The checklist should begin with a brief summary statement of its objective, the recommended frequency of implementation, and the number and required qualifications of the personnel completing the checklist.

The level of detail within a specific checklist should be directly related to the level of complexity of the system being checked. Requirements should be structured to identify problems to be flagged for further attention, and to ensure that standards, industry practices, and procedures are being followed.

Once drafted, the checklist should be independently reviewed by other individuals also familiar with its objective. Finally, the preparer and reviewers should walk through the checklist (i.e., physically check it against each checklist subject) to confirm that appropriate attributes have not been omitted from the list. Occasionally the order of a checklist will be reversed during the confirmation process. Reverse-order confirmation helps to identify steps or attributes that are out of order or omitted during development. Reverse-order confirmation is typically used for checklists that evaluate complex systems.

The checklists should be used and maintained during the life of a project, including decommissioning. This requires periodic evaluation of the checklist to ensure its integrity and accuracy. Checklists should be updated after each modification is implemented and after every major outage when equipment is replaced or modified substantially.

SUPPORTING DOCUMENTATION

The basis for including an item on a checklist is whether it contributes to the checklist objective. The checklist objective, therefore, dictates the type of supporting documentation necessary for its development. Checklists may be created for a variety of purposes as well as for various stages of a project's life cycle, and the documentation in support of checklist development will vary from list to list. The following recommendations are provided for identifying attributes when developing plant operation or design checklists:

- Plant Operations: Operating procedures, including startup and shutdown; inspection and maintenance procedures; operator training manual; system drawings, including support and shared system drawings; vendor specifications and manuals; plot plan; equipment performance history; controls description; emergency or upset procedures; problem reports.
- Plant Design: Piping and instrumentation diagrams (P&IDs), process flow diagram (PFDs), electrical one-line diagrams, control logic diagrams, mass and energy balances, fire and electrical area classification drawings, plot plan, criteria for design and operation, criteria for establishing design temperatures pressures or materials of construction, equipment arrangement drawings, purchase specifications, vendor manuals, consensus standards (e.g., practices related to special handling, toxicity, flammability, corrosiveness, or reactivity).

METHODOLOGY IMPLEMENTATION

Checklists can be applied during all stages of a project's life. During conceptual engineering and design development, checklists help to identify potential hazards, safe designs, and areas that may require further study. Construction checklists are an effective means of verifying quality assurance, testing, and conformity to established design standards. Initial, or startup, operation checklists help the operator align systems for commercial operation and test equipment performance. These checklists address, when necessary, equipment or system conditioning for normal operation, such as system evacuation, system purging and drying, installation of blinds, or car seals on valves. Operations checklists ensure that the facility is being operated according to procedures and that personnel are performing on-line calibration, testing standby systems, and verifying system alignment (important for batch operations). Standby or shutdown checklists verify equipment status prior to removal for maintenance, especially when hazardous operations are required. They can also ensure availability of alternate systems, which is important for continuous processes where on-line maintenance may be performed, and provide a cross-reference to the appropriate maintenance-procedure checklist.

While checklists must be prepared by experienced staff, they may be implemented by less experienced personnel who will require training sufficient to allow them to complete the checklist.

RESULTS

Checklist results are qualitative in nature. They provide insights into the degree of compliance with prescribed procedures and identify potential hazards. Checklists may also provide a status for each operation (e.g., verification of system operating conditions or component operating status).

Checklists do not provide any insight into systems interactions; they merely provide the status of the item in question. For example, a checklist attribute may be "Pump A Running?" Answer: "Yes" or "No." The checklist indicates whether or not the field pump is running, but it cannot provide any insight into the reason for pump failure.

ADVANTAGES AND DISADVANTAGES

A checklist is the simplest hazard identification technique available. Although checklist development requires knowledgeable personnel, relatively untrained personnel can use them effectively if they have been adequately instructed in the checklist's application. The use of a checklist is a rule-based behavior, meaning that someone follows written instructions to perform a particular act each time it is required. So implementation requires only minimal training.

A checklist provides simple documentation of the status of an item but does not prioritize items that have been labeled as unacceptable. A checklist can focus only on a single item at a time, so it cannot identify hazards as a result of interdependencies or interactions between processes or procedures. Checklists also tend to limit focus to the immediate attribute, restricting personnel from using their imagination or intuition to identify potential problem areas. The most significant disadvantage of a checklist is that it is only as good as the ability and prior experience of the preparer. A checklist prepared by an inexperienced or unqualified person might neglect critical items that could identify hazards worthy of further study. Finally, if checklists are not independently verified, there may be a potential for omitted items to go undetected.

EXAMPLE CHECKLISTS

The following sample checklists have been developed to assist a hazards analyst in identifying problems that may require further attention. The examples are general, and therefore a paragraph stating the objective and describing the focus for the checklist is not provided. The checklists should be modified to reflect specific objectives and facilities prior to application in an actual facility.

Checklist A—Plant Organization and Administration

1. Organization
   a. Corporate organization chart detailing areas of responsibility for each division and the name and telephone number of the key person responsible.
   b. Divisional organization chart identifying supervisors, group assignments and functions, and the names of personnel in each group.
   c. Is a procedure in place to periodically update these charts and distribute to appropriate personnel?
   d. Specialty areas highlighted for quick reference (e.g., Fire Warden, Plant Safety Supervisor, Emergency Response Coordinator).
   e. Are adequate facilities available (e.g., offices, technical library, warehouses, laboratories)?

    f. Are personnel with technical expertise readily available?

    g. Are there any plans for expansion or modernization of the facility?

2. Administration
   a. Plant Operators
      1) Are plant procedures readily available?
      2) Are emergency procedures available?
      3) Are the operators periodically evaluated to check their competency?
      4) Are operators periodically retrained?
      5) Has the training program been formalized?
      6) Are the operators periodically drilled on responses to random simulated emergency situations?
   b. Maintenance Group
      1) Are adequate facilities available (e.g., offices, records library, warehouses, maintenance equipment)?
      2) Are vendor equipment manuals available for quick reference?
      3) Have personnel been periodically retrained and educated on new techniques?
      4) Are personnel supported by an engineering staff or contracted maintenance professionals?
      5) Is a program in place for preventive and predictive maintenance?
      6) Are findings from maintenance activities cataloged and routed to the engineering staff for evaluation?
      7) Are functions and responsibilities, especially safety and inspection interfaces, well defined?
   c. Emergency Response Group
      1) How is the plant shut down in case of a fire emergency?
         a) Panic button to emergency shut-down (ESD) system.
         b) Individual motor-operated valves (MOVs).
         c) Fire alarm to ESD system.
         d) Manual valve operation.
      2) Is an emergency response plan available and supported by management?
      3) Are procedures in place for activation of the plan in place?
      4) Emergency protocol: Is there a notification sequence, and is it prominently displayed on the operating floor and in the control room?
      5) Is the plan evaluated and updated periodically?

6) Have local authorities been briefed and trained in the plan and its major features?
7) Is emergency support equipment in place and adequately maintained?
8) Are procedures for deactivation and recovery detailed in the plan?

Checklist B—General Operations

1. Inventory Control
   a. Are dangerous or hazardous substances stored in remote locations?
   b. Is on-site inventory maintained at a minimum acceptable level?
   c. Are detectors and alarms provided for detection of leaks or spills?
   d. Is inventory maintained in a safe fashion (e.g., are drums stacked a maximum of two high) and hazardous substances segregated?
   e. Is storage area in compliance with local building codes (e.g., electrical utilities, fire protection)?

2. Production Area
   a. Are dangerous or hazardous substances staged to the process in an acceptable manner?
   b. Is staging area protected from adjacent operations or traffic?
   c. Has process instrumentation been adequately maintained?
   d. Is local instrumentation readily accessible or visible to operators from local control panels?
   e. Are drain connections valved and capped?
   f. Are maintenance valves locked in the appropriate position for operation?
   g. Are local annunciators furnished to alert floor operators of problems?

3. Intermediaries and By-product Discharges
   a. Are all hazardous intermediaries properly labeled?
   b. Are discharges monitored?
   c. Are safeguards in place to prevent improper discharges?
   d. Are vents routed to flares or scrubbers?

4. Final Product Handling
   a. Is product packaged for on-site use or for off-site use?
   b. Is product adequately protected from other operations?
   c. Is product adequately labeled?

5. Control Room Operations and Communications
   a. Is a working communication system available to floor operations?
   b. Are shift logs maintained? Do they document plant alarms, upset conditions, and trips? Are log sheets or automatic data logging used?
   c. Does management review shift logs to evaluate operators and operations performance?
   d. Are trainees closely supervised by certified operators?
   e. Is a procedure in place for addressing operator concerns?
   f. Has the operations manual been marked up by the operator to reflect actual conditions?
   g. Are maintenance personnel notified immediately of failures on the floor?

6. Documentation
   a. Are plant drawings maintained in accordance with actual plant conditions?
   b. When drawings are updated, is a procedure in place to update all controlled copies?
   c. Is there a procedure to control distribution of drawings and other procedures?
   d. Are appropriate groups assigned responsibilities for updating and maintaining procedures?
   e. Operations Manuals
       1) Is supervisory manual clear and concise?
       2) Is supervisory manual self-contained, or does it cross-reference vendors' instructions?
       3) Are vendor's instructions readily available?
       4) Does manual contain
           a) Process flow diagrams (PFD)?
           b) Piping and instrumentation diagrams (P&ID)?
           c) Alarm set points?
           d) Pressure safety valve (PSV) set points?
           e) Sampling techniques?
           f) Product specifications?
           g) Laboratory test requirements?
   f. Operating Instructions
       1) Do valve-by-valve instructions exist? Who wrote them?
       2) Are the instructions used by trainee operators?
       3) Are the instructions updated based on experience and plant changes?

g. Shift Turnover Book
  1) Is it reviewed by plant management?
  2) Are frequent alarms logged?
  3) Are all trips documented in the log? Does Operator give explanation?

7. Plant Operations
  a. What is the plant on-stream factor?
  b. What are the most frequent reasons for unscheduled shutdowns?
    1) Power supply interruption.
    2) Utility failure
      a) Steam supply.
      b) Cooling water.
      c) Instrument air supply.
    3) Major equipment failure.
    4) Furnace problems or failure.
    5) Interlock system induced shutdown.
    6) Operator error.
  c. Which area or areas of the plant require the most frequent operator attention?
  d. What is the personnel turnover rate in various departments?
    1) Operating.
    2) Maintenance.
  e. What type of shift change handover is practiced for operators?
    1) On-the-job relief.
    2) In control room relief.
    3) At plant gate relief.
    4) Verbal handover only.
    5) Verbal and written handover.
  f. Are operators being rotated to different areas of the plant on a regular basis?
  g. Are new operators being trained in the plant?
  h. Are operators being evaluated for competency on a regular basis?

8. Training
  a. Is a full-time training program in effect?
    1) Have operator proficiency and qualification requirements been identified and adhered to?
    2) Are oral examinations performed?
    3) Are written examinations performed?
    4) Are emergency simulations enacted?

5) Are alternate operating modes discussed and researched?
6) Are equipment qualifications reviewed with operators?
7) Are interim training sessions held when plant modifications are performed?
8) Is a full-time training instructor assigned for process operators and maintenance personnel?
9) Is a training room available with various visual aid apparatus (e.g., overhead projector, video recorder/monitor, large drawings and charts, film projector)?
10) Is a training course curriculum available with printed handbooks, test sheets, and other learning aids?
11) Are process operators and maintenance personnel kept up to date when plant modifications or new equipment are introduced by retraining?

Checklist C—Maintenance

1. Has a maintenance program been formalized?
    a. Are warehouse inventory control procedures in place?
    b. Is an automated or manual inventory procurement program in place?
    c. Can a surplus of hazardous materials be procured?

2. How are maintenance department activities coordinated with plant operating?

3. Are maintenance personnel available when required by operations?

4. Is equipment usually operated at its optimum design range? If not, what problems have been encountered?

5. Has degraded equipment forced operating requirements to be outside design parameters?
    a. Is the instrumentation and control system maintained adequately?

6. Is operation of instrumentation in the manual mode required because of
    a. Process stability problems?
    b. Inadequate maintenance?

7. Are analyses performed to determine the best approach:

   a. Repair/delay.
   b. Repair/replace.

8. Who determines repair or replacement?

9. What efforts are made to upgrade equipment?

10. How are feedback and new technology incorporated?

11. Are spare parts available in support of maintenance? Which spare parts are fabricated at facility? Are all spare parts original equipment by manufacturer? Is inventory inspected periodically?

12. Are spare parts and chemical stocks replaced after maintenance? How are stocking levels determined? Is a spare part inventory available?

13. What type of storage system exists? Are new materials inspected?

14. Are spare parts and chemical inventories interfaced with other plants?

15. Are replacement materials made in kind or is the state of the art considered? Is obsolescence considered?

16. Are spare parts available for maintenance during an unscheduled shutdown?

17. Are spares and materials classified by replacement cost, frequency, delivery, labor intensity, sources, or effect on production or safety?

18. What records are maintained?
   a. Time and personnel staffing records.
   b. Equipment and machinery maintenance logs.
   c. Record system (coding and inventory control).
   d. Lubrication schedules.
   e. Instrument and control calibration.
   f. Actual expenditures and schedules vs. budgets (performance).
   g. Frequency of unscheduled shutdowns and causes.
   h. Are maintenance findings routed to the engineering staff for evaluations?

19. Technical manuals and prints.
   a. Are vendors' manuals available and up to date?

    b. Are prints available and up to date?

    c. Are as-built drawings up to date?

    d. Are vendor recommendations followed?

20. Are written maintenance orders or work requests used and is there a written procedure defining the system?

21. Do work requests contain the following information?
    a. Clear description of malfunction or problem.
    b. Description of work.
    c. Tools required and special test equipment.
    d. Tagging requirements.
    e. Test required.
    f. Safety precautions.
    g. Drawings or procedures' references.
    h. Identification of material needed and spare parts.
    i. Priority (who assigns it?).
    j. Estimated time to repair.
    k. Status of plant during repair.
    l. Personnel requirements.
    m. Means for documenting cost.
    n. Approval and authorization provisions.

22. Are sparkproof tools available? Who determines whether sparkproof tools are to be used?

23. Work schedules: Are the following used?
    a. Maintenance staff available for all shifts.
    b. Daily and weekly work schedules.
    c. Personnel assignments.
    d. Long-range planning schedules.

24. Are job planners used?

25. Are maintenance schedules coordinated with plant operation?

26. Who coordinates the turnaround?

27. What meetings, if any, are held during turnaround?

28. Is the sequence of maintenance work defined?
    If so, are the functions of each step in the procedure defined (e.g., job planner, coordinator)?

29. Is there a preventive maintenance program?

30. Turnaround planning.
   a. Is planning process a daily activity? How is backlog addressed?
   b. Are priorities established for modifications or repairs during an unscheduled plant shutdown?
   c. What is the constraint to reducing typical scheduled turnaround time?
   d. How is the interface of area activities with systems activities achieved?

31. Personnel.
   a. Morale.
      1) Has impact of daily work on quality of life been stressed?
   b. Overtime practices
      1) Which department shows the highest amount of overtime?
   c. Use of subcontractors?
      1) For routine maintenance.
      2) For specialty services.
      3) For plant turnaround.

32. Training.
   a. Training records.
   b. Apprentice training or similar program.
   c. Periodic review training.
   d. Vendor schools.
   e. On-the-Job training.
   f. Personnel goals.
   g. Levels of qualification.
   h. Educational and training material available.
   i. Does management support the training effort?
      1) Organizationally.
      2) With budget and resources.

The following checklist was developed to verify various activities performed during a modification.

Checklist D—Inspection

1. Replacement Equipment Procurement
   a. Are appropriate specifications prepared? Have data sheets been completed and verified?
      1) Are references to consensus standards included?

    b. Have vendor shops been visited to verify qualifications?
       1) Is a quality-assurance program in place?
       2) Is a certification program available?
    c. Is a receipt inspection program in place?
       1) Verification against procurement specifications required?

2. Equipment Storage
    a. Have appropriate provisions and precautions been taken to protect equipment while it is in storage?
    b. Has shelf life of subcomponents been noted?
    c. Is equipment protected from other storage area activity?

3. Piping and Vessels
    a. Is ultrasonic thickness testing of vessels and piping done on a regular basis (e.g., during turnaround)?
    b. What other methods of inspection and nondestructive testing are used (e.g., dye penetrant, magnetic particle)?
    c. Does the maintenance department do this testing or are there special personnel for inspection and testing? Is new or modified piping tested, and how is this done?
    d. How often and in what manner is PSV testing performed?
    e. Are corrosion-prone areas of process piping and vessels inspected on a regular basis?
    f. If pipe metal failure or weld failure has occurred, was analysis done by outside laboratories?
    g. Is X-ray inspection apparatus available; can plant maintenance personnel interpret X-rays?

4. Instrumentation
    a. Are trip circuits tested on a regular basis?
       1) Are procedures prepared for this work?
       2) Is there a sign-off list for these tests?
       3) Are operators doing a functional test after each trip to verify system availability?
       4) Are bypass switches provided for testing?
       5) Are these bypass switches accessible for all personnel or are they locked in a cabinet, with special personnel responsible for keys?
    b. Are instruments zero checked or calibrated on a routine basis, or are they checked when reason for accuracy or doubt exists?
    c. Is an instrument technician available on a 24-hour-per-day basis?

      1) Are instrument technicians on call (Is a roster of personnel available)?

      2) Are instrument technicians' skills upgraded on a routine basis through special training or other means?

5. Pumps and Compressors

   a. Are records kept to trace frequency of failure of seals and other parts? Do records include exact description of spares used, mechanics who did job, and other job specifics?

   b. Are compressors or other large, nonspared machinery inspected on routine basis (such as during turnaround), or is maintenance based on problem observation?

      1) Is large rotating machinery fitted with vibration-analysis equipment?

      2) Is portable vibration equipment available for spot-checks?

      3) Is vibration spot-checking done on a regular basis?

      4) Was large rotating machinery voice-printed for vibration at initial plant startup?

   c. Is major overhaul performed by plant maintenance, or are vendors' representatives called in?

      1) Is this work done by an outside contractor or shop?

      2) What is experience with outside shop work, if any?

Checklist E—Safety

1. Are procedures available and used when isolating equipment for maintenance?

2. Is Safety Department responsible for work order signature, or is this done by operations or maintenance personnel?

3. Are blind lists made for each isolation job, who keeps them, and who checks that all are installed or removed?

4. Is safety and life-saving equipment inspected on a regular basis, and who is responsible for this work?

5. Are operators and maintenance personnel instructed and trained in firefighting and first-aid procedures?

6. Are plant personnel trained to respond to major emergency situations?

7. What is the level of firefighting equipment or capability in the plant? Is outside backup available?

8. Is emergency medical treatment available at all times?

9. Is an automatic gas or vapor detection system installed showing location and alarm point in control room?

10. Is the fire water system tested on a regular basis?

11. Are steam or water curtains provided for critical equipment and areas?

12. Are automatic fire-extinguishing systems installed (Halon, CO, Foam, etc.)?

13. Is the control room located and built to withstand certain fire and explosion hazards?

14. Are remotely operated emergency shutoff valves provided? If so, are these tested on a regular basis?

15. Are air packs provided; if so, what is their location and who tests and refills these? What are site rules regarding personnel with beards?

16. How are vessels checked before entering? What nitrogen safety procedures are used?

17. How are vessels freed of hydrocarbons and mercury before entering? How are they checked?

18. Is safety consciousness emphasized?

19. Are good safety records rewarded in any way?

20. Is a safety committee established in the Operations Department? In the Maintenance Department?

21. Are standard operating procedures reviewed for safety hazards? Who reviews them?

22. Is the Safety Department entitled to enforce housekeeping?

23. Which department is responsible for gate perimeter security?

24. Is all safety equipment checked on a regular basis for proper function? Who signs off?

25. Is safety shoe and eyeglass protection mandatory?

26. Are lines marked for contents (acid, caustic substances, etc.)? Are adequate safety showers and eyewash facilities provided?

27. Is a safety training course in effect? How often does it convene, who takes part, who teaches it? How many hours per month are spent in training?

28. Are operating and maintenance techniques updated when new equipment is introduced?

29. Are motors, switch panels, ignition panels, and solenoids adequate for the electrical area classification?

30. Is the integrity of electrical grounds maintained?

31. Are fire isolation considerations applied to curbs, drains, or sewer systems?

32. Are operating personnel instructed in purpose and functioning of mechanical safety devices (e.g., tank breathers, overspeed protective devices, float switches, trip systems)?

33. Are charts available identifying every chemical or compound being used in the plant, and are toxicity and first-aid measures described?

34. Are ignition sources (switchgear, smoking areas, workshops, etc.) close to the boundary of a hazardous area?

Checklist F—HAZOPS

1. Is there a hazards and operability study available for facilities?

2. Is each piece of equipment protected against overpressure caused by operational upsets?

3. Is each piece of equipment protected against overpressure caused by fire?

4. What coincidental conditions is the flare system designed for?

5. Can PSVs be taken out of service when the plant is on-line?

6. Have any modifications been made since the plant was built? If so, how are the modifications documented? Is the HAZOPS study updated? Are as-built drawings updated?

7. Is it possible to overpressurize atmospheric storage tanks by
   a. Loss of liquid level in vessel feeding tanks?
   b. High vapor pressure material being sent to tanks?

8. Are trip circuits normally energized or normally deenergized?

9. How are trip circuits tested, and how often?

10. What are consequences of trip failure?

11. What are the consequences of temporary fuel gas failure? Can gas be restored to a hot furnace?

12. Is rotating machinery protected against backspin when a relief valve blows?

13. Is the flare system protected against liquid entrainment?

14. What is the design velocity at flare tip?

15. What is the radiation level at the edge of the flare field? Is the flare field fenced off?

16. What is the location of the oily sewer relative to forced draft fans and other combustion sources?

17. Are combustible gas detectors installed at all combustion sources?

18. What trips are bypassed in day-to-day operation? How are they documented?

19. How does the plant operate compared to design:
    a. Closer to PSV settings?
    b. Higher throughout?
    c. Colder?
    d. Hotter?
    e. Lower voltage?
    f. High cooling water?

SUGGESTED READINGS

Balemans, A. W. M., et al. 1974. *Check-list: Guidelines for safe design of process plants*. Paper read at First International Loss Prevention Symposium.

Hettig, S. G. 1966. A project checklist of safety hazards. *Chemical Engineering* 73(26).

King, R., and Magid, J. 1979. *Industrial Hazard and Safety Handbook*. London: Newnes-Butterworths.

Marinak, M. J. 1967. Pilot plant prestart safety checklist. *Chemical Engineering Progress* 63(11).

4

Preliminary Hazards Analysis

Robert T. Hessian, Jr. and Jack N. Rubin

Preliminary hazards analysis is a hazard identification technique that focuses on the conceptual design phase of a project. The purpose of this analysis is to identify early in the design process the potential hazards associated with, or inherent in, a process design, thus eliminating costly and time-consuming delays caused by design changes made later. With the use of this technique, the areas of the design critical to safety and the potential hazards can be addressed by incorporating safety factors into the design criteria.

DESCRIPTION OF METHODOLOGY

An assessment of the conceptual design should be conducted for the purpose of identifying and examining hazards related to feedstock materials, major process components, utility and support systems, environmental factors, proposed operations, facilities, and safeguards.

Feedstock

A feedstock review focuses on all aspects of handling, from on-site arrival to use in the process. Identified potential hazards provide guidance for the development of administrative controls and other criteria for operation.

Physical properties are reviewed for compatibility with the proposed materials of construction. Mixtures may be investigated to determine whether any potentially explosive, flammable, reactive, or toxic threshold limits exist. This same review should also be extended to include all process intermediates, chemical additives, and products.

Major Process Components

The purpose of a major components review is to check the normal, intermittent, and upset conditions and their relationship to the proposed design criteria, to determine whether appropriate equipment design conditions have been specified. Preliminary or scoping calculations are performed to determine the maximum system parameters (e.g., temperature, pressure, concentration, pH, or time), for both continuous and intermittent operating conditions, so that appropriate safety factors are developed and incorporated into the design criteria. It is especially important to establish criteria for startup, shutdown, and intermittent operations.

Utility and Support Systems

Support systems are by definition those auxiliary systems that assist major process components in performing their intended function (instrument air, steam, cooling water, etc.). Thus, the reliability of major process components is directly dependent upon their support and utility systems. These systems should, therefore, be reviewed to determine whether their failure or partial failure could create a hazardous condition in the process system. Appropriate design criteria for the support system, identified as a result of this review, help ensure an acceptable level of reliability within the systems and directly enhance process reliability and safety. Failure of the cooling water or electrical system can be particularly dangerous in many facilities.

Interconnecting systems are typically found in processes that are designed to operate on a continuous basis. These systems frequently require additives or other compounds necessary for the main process. The intermediaries either maintain the reactivity within acceptable ranges or promote catalyst selectivity. Hydrogen partial pressure is critical to most hydrogenation processes and may affect the mechanical design of the unit.

Interconnecting systems should be reviewed to determine whether accidental, improper, or off-specification additives may be introduced into the main process stream and, if so, whether a hazardous condition could result. Design criteria to control critical intermediary additives to the process should be identified and incorporated into the process design specifications.

Environmental Factors

The ambient environment should be reviewed to determine whether any external events may create potential hazards for the process or process design. External sources of vibration, elevated or very low winter temperatures, and fire or corrosive substances are a few of the potential hazards that should be considered when reviewing a process area. Processes located outdoors should consider direct exposure to adverse meteorological conditions and wind direction. Potential hazards caused by environmental factors should be considered in the design (as with adverse meteorological conditions for outdoor facilities) or addressed with the use of mitigative systems, such as potential fire sources warranting the use of sprinklers. Some states, such as California, and certain federal regulations require additional safety factors to protect against naturally occurring events such as earthquakes.

Proposed Operations

A review of the proposed operations for the process should be conducted. This review is done to identify potential hazards that may be created during the various phases of the process or operating modes, including startup, emergency scenarios, maintenance, testing, and shutdown. In addition, the plot plan should be reviewed to identify potential hazards caused by deficiencies in access or egress, and the need for life-support equipment in potentially high-risk areas.

Since this technique is applied relatively early in the project design cycle, operating procedures, maintenance requirements, equipment testing cycles, and emergency operating procedures may not yet be available. If procedures and precautions have been previously developed for similar systems, they may be used in the review of the new system. When a totally new design concept is being developed and similar designs do not already exist, individual equipment vendor data for equipment with a previous service history with similar chemicals or reactions may be used. Experienced engineers can offer valuable insight into potentially dangerous situations.

This review will identify potential hazards associated with administrative controls, including but not limited to operator errors, ergonomics (i.e., the study of the human-to-technological interface), and potentially harsh environmental conditions where operators must be present during upset or transient conditions.

Facilities

A review of system facilities should be conducted for sources of potential hazards, not only to the process but to the operators as well. Critical items to be included in the review are high-energy piping systems, electrical power sources, other utilities, drainage from the process area, waste processing, facility maintenance requirements, lighting, ventilation, and area monitoring. Each system has the potential to cause damage to equipment or to the environment or may be dangerous to personnel.

Safeguards

A safeguards review examines the adequacy of proposed safety equipment (e.g., interlocks, pressure-relief devices, system redundancy) and mitigative systems. Safety components are identified as those components that control or contain system reactions during transient or upset conditions. Mitigative systems are those systems added to the process or facility to mitigate the effects of predetermined adverse conditions (such as safety valves) that cannot be contained by the process. The mitigative system review may also include a comparison between alternative technologies, such as water sprays versus scrubbers.

A final assessment of the safety of a new design should be made by comparing the proposed process to alternative configurations when possible. This final assessment is a comparison of inherent risks between the chosen concept and alternative configurations. While inherent risks alone are not the basis for deciding process viability, they can provide insight into the minimum achievable risk associated with a given design.

SUPPORTING DOCUMENTATION

Since a preliminary hazards analysis is conducted early in a project's life, a limited number of plant-specific documents are available to support the analysis. Available data are found in the conceptual design documents, including process flow diagrams, mass balances, preliminary equipment sizing calculations, laboratory experiments for determination of reactivity limits (e.g., lower explosive limits), summaries of intermediary and final reactions, projected reaction durations, and listings of raw materials and their approximate inventories.

There is no better teacher than experience. Therefore, historical data should be obtained and, if available, so should data from sister plants or plants of similar design. Where the process concept is one of a kind or new technology, historical data should be obtained from other processes that have employed similar equipment from equipment vendors or process specialists.

METHODOLOGY IMPLEMENTATION

The purpose of the preliminary hazards analysis is to identify hazards within each category described in the preceding methodology section. The intent of the analysis is to review the process from the introduction of raw materials on-site, through final product generation or packaging.

The technique should be started as soon as the basic concepts for the project are established and the process flow diagrams and mass balances have been developed. The analysis may be conducted by a single analyst or by a team of analysts. Complex processes that involve multiple stages of reaction, require multiple support systems, or are protected by several safeguard systems should be reviewed by a multidisciplinary team. The team may also call upon additional expertise for specialized areas, such as fire protection or advanced control.

RESULTS

The findings of the analysis will be a list of acceptable safety-related design features and additional features that may be added to the project to improve safety. These results may include precautions or operator safeguards to be incorporated into the final administrative controls, recommended modifications to the process, enhancements to the design of associated facilities, and desirable alternatives that would significantly reduce or limit inherent hazards.

Results from this analysis will be qualitative in nature. The findings will not indicate the priority or relative importance that each hazard presents, and each finding is given equal consideration.

ADVANTAGES AND DISADVANTAGES

This technique, applied early in the project life cycle, helps to eliminate hazards and, thus, to avoid costly design modifications later. Since the project has only preliminary or conceptual design documents available, the analysis requires a relatively small effort compared to other hazard identification techniques. This analysis fortifies the proposed process design by incorporating additional safety factors into the design criteria.

The disadvantage of this technique is that it does not allow for prioritization of its findings, and, as a result, each hazard and remedial recommendation is given equal consideration. In addition, it is not as systematic as other techniques (e.g., HAZOP) and relies heavily on the process experience, intuition, and imagination of the analyst.

EXAMPLE PROBLEM

The process shown in Figure 4-1 utilizes hydrofluoric acid (HF) as a catalyst. This process flow diagram depicts how the chemical is added to and used by the process. Since this is a new process concept with no track record, the analyst must search for data for equipment that has been previously used for HF service in other processes.

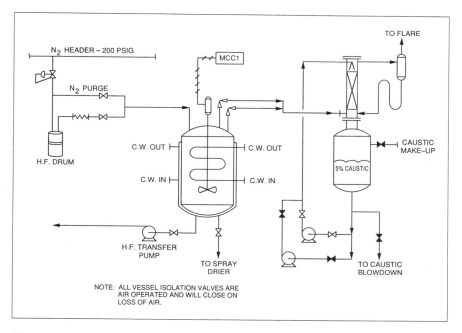

Figure 4-1 Example flow diagram. All vessel isolation valves are air operated and will close on loss of air.

The analysts realize that HF released to the atmosphere is acutely toxic and identify its potential release as the objective of this analysis. The analysts now review each process category to determine the possible causes of HF releases.

FEEDSTOCKS

- The HF addition area is not sheltered or protected from other traffic on the process floor, leading to potential rupture of shipping or storage drums.
- The transfer hose between the storage drum and process may leak or fail, releasing HF to the environment.
- Nitrogen is used to pressure HF into the process. A failure of the N_2 pressure regulator may overpressurize the storage drum, releasing HF to the environment.
- HF is a highly corrosive material. If the raw material feed line to the reactor has a high relative humidity, corrosion of the pipe may be accelerated. This creates a situation in which the reduced wall thick-

ness might fail during raw material transfer, releasing HF to the environment.

MAJOR PROCESS COMPONENTS

• The exothermic reaction is cooled by a plant cooling water system. Loss of the cooling water causes a runaway reaction, with the evolution of gas. The excess pressure generated lifts the vessel relief valves, venting flow to the flare. Sizing of the relief valves is critical to safety. Inadequately sized valves may cause an overpressure in the vessel that may rupture flange gaskets or the vessel itself.

• The reactor is equipped with a single drive agitator. The agitator has a single electrical source without backup. Failure of the drive motor or the power source results in a separation of the reactants. Since the reactants need to be mixed for the reaction to occur, loss of the agitator results in a loss of the reaction and generates an off-specification product. Although this failure does not pose a hazard, plant designers should consider the merits of an alternate power source to improve production reliability.

• When the cake has formed, spent HF is separated from the reaction at the end of the batch process. Spent HF is pumped to the acid regenerators. The preliminary process design does not indicate whether pump seals are of the mechanical type, with provisions for being flushed out during normal operation. Failure to flush the pump seals leads to seal failure and release of HF to the environment. Selection of a leak-proof pump might also be considered for this service.

UTILITY AND SUPPORT SYSTEMS

• Cooling water system includes three 50-percent-capacity centrifugal pumps. These pumps are initially indicated as manually controlled. The preliminary documentation provides for local flow meters to be located in the pump house. Since the process is dependent upon the cooling system for removing excess heat generated by the exotherm (as through a reaction), remote indication of system performance for operators should be provided. Remote startup and control of the third pump should also be provided, since loss of cooling can have an adverse affect on the reactor and process-related equipment if one of the two on-line pumps fails. An alternative consideration would be a third pump, driven by a steam turbine, with an autostart system activated by low pressure.

• The instrument air system has been configured with only one com-

PROJECT: _____ CLIENT: _____

SYSTEM: _____ DATE: _____

PHA CATEGORY	DESCRIPTION OF HAZARD	RECOMMENDATION	RESOLUTION

Figure 4-2 Preliminary hazards analysis table.

pressor. A second compressor should be considered to improve the reliability of the air system. An alternative to a second compressor is the use of accumulators on valves and equipment that must perform a desired safety function, regardless of the availability of the air supply.

ENVIRONMENTAL FACTORS

- Because of the highly reactive nature of the chemicals used in this reaction, controls on the fire-protection system should be calibrated so that all other fire-protection measures are implemented before initiating sprinkler sprays.

PROPOSED OPERATIONS

- A review of the proposed operations indicates that every time an operator decides to take a process grab sample, there will be a certain amount of fugitive releases to the atmosphere, exposing the operator to some hazardous vapors. The sampling method should be revised so that the sampling process contains all vapors, thus eliminating fugitive releases in the presence of the operator.

This example problem could be configured in a tabular format as well. The benefit of a tabular format is that provisions can be made on the table for tracking subsequent actions taken relative to the identified concern. An example of a tabular format is shown in Figure 4-2.

SUGGESTED READINGS

Gid, J. 1979. *Industrial Hazard and Safety Handbook.* London: Newnes-Butterworths.

Marinak, M. J. 1967. Pilot plant prestart safety checklist. *Chemical Engineering Progress* 63(11).

5

Safety Audits

Robert T. Hessian, Jr. and Jack N. Rubin

The safety audit is a procedure by which a plant or process is inspected. An auditor or audit team reviews critical plant features to verify the implementation and effectiveness of appropriate design criteria, operating conditions and procedures, safety measures, and related risk management programs. The audit also includes interviews with key individuals, including supervisory, operations, maintenance, and safety personnel.

Safety audits should be performed periodically as a means of certifying that the plant's safety program is in place and is being implemented effectively. Periodic reviews also reflect the program's performance since the last review and identify areas where there may have been a lapse in procedures or dedication of effort.

Safety audits are typically performed in conjunction with other hazard identification techniques, such as WHAT-IF analysis, failure modes and effects analysis, or hazards and operability studies, which are discussed in Chapters 6, 7, and 8, respectively.

This chapter provides guidance on how to conduct a comprehensive safety audit for a new or existing facility, describes how these procedures are typically developed and maintained, and gives recommendations and examples for documenting the findings of the safety audit.

DESCRIPTION OF METHODOLOGY

Safety Audit: New Facility

In general, new facilities are designed to take advantage of modern commercially demonstrated technology and available process enhancements. A comprehensive safety audit for a new facility requires a full review of the facility's design criteria. The following is a typical list of required design information:

- Basis of design

- Process description
- Process flow diagrams (PFDs)
- Equipment data sheets
- Piping and instrumentation diagrams (P&IDs)
- Equipment arrangement drawings
- Site plan
- Site location drawing
- Electrical one-line diagrams
- Electrical area classification drawings
- Specifications for piping
- Specifications for instrumentation and controls
- Specifications for safety relief devices.

In addition to the information on technology, standard operating procedures, and maintenance procedures, information on training and emergency preparedness should be provided. This additional documentation provides insight into how well the plant is operated and maintained.

Safety Audit: Existing Facility

Most processing facilities are not new, and many were designed and built to standards that might not be acceptable today. Some systems, such as instrumentation, can be updated and replaced. Whatever the age of the plant, one of the main functions of a safety audit is to assist in upgrading existing facilities to an acceptable level of safety.

A safety audit of an existing facility, therefore, involves visual inspections and reviews of actual plant configuration and operating conditions. Comparisons should be made with the design documentation, including any amendments that have occurred since the documents were originally developed.

A plant inspection, while carrying the piping and instrumentation diagrams, can be very useful in determining the validity of the drawings and the accessibility of instrumentation and valves. Comparisons with design specifications and procedures determine whether the plant emergency and mitigative systems are functional, whether equipment is being maintained in accordance with manufacturers' recommendations, and whether instrumentation and controls are functioning in accordance with their intended service.

A comprehensive safety audit requires interviews with plant operators and maintenance crews, and a review of plant operating and maintenance logs. This helps to determine whether the facility is being operated within its design criteria and established administrative controls. Interviews are also important as a means of involving plant personnel, who are generally very cooperative and interested in making their plant as safe as possible.

SUPPORTING DOCUMENTATION

In addition to the plant's criteria for design and operation with amendments, the audit team needs current revisions of all other plant documents within the scope of the current safety review. These include plant operating and maintenance logs, standard operating procedures, training manuals, and emergency preparedness programs.

METHODOLOGY IMPLEMENTATION

A safety audit may be performed by one person or a team, depending on the complexity of the plant or the process to be audited. In general, large plants should be reviewed by a multidisciplinary team. To ensure objectivity, the team should include as few plant personnel as possible who are involved in the day-to-day operations of the area scheduled for audit. All relevant plant data should be gathered and routed to all team members for review and familiarization before the start of the audit. Each team member should then identify appropriate plant personnel to be interviewed as a part of the audit.

Determining which plant personnel are to be interviewed requires evaluation of all areas of the process scheduled for audit. These include process equipment, instrumentation, maintenance, support systems, utilities, mitigative systems, fire protection, life-support systems or equipment, safety precautions and training, and security against operator errors or sabotage.

Upon completion of all interviews, the team should conduct a walk-through of the plant to identify deficiencies, deviations from standard procedures, or failures of the administrative controls, design criteria, and operating goals. Drawings should be compared with the physical arrangement to verify that current documentation is consistent with the plant. The team should evaluate plant records, including operator logs, to identify process areas where the number of personnel injuries may have increased and to look for sudden drops in personnel injuries to determine whether changes may be applicable elsewhere. Maintenance records should be reviewed for indications of age or accelerated wear of the process equipment. Equipment calibration reports should be checked to identify troublesome components that may require additional attention or replacement. Safety device-system test results should be examined for verification of capacity, function, and availability.

REPORTING RESULTS

The final task of any audit is a report that documents all findings and impressions. Positive indications of safe operations should be reported. Recommendations addressing hazardous conditions should be supported by appropriate justification or references.

The safety audit report provides corporate management with an overview of the plant and process. It helps to identify the level of performance for various aspects of operations and can identify where the focus of future inspections or audits should be shifted. Comprehensive safety audits also closely examine prior focuses of attention to determine whether previous modifications were justified.

One purpose of a safety audit is to verify that the existing facilities and associated programs are in place and functioning to provide the greatest possible degree of safety. Therefore, when the team is reporting its findings, the primary concern is not the cost associated with implementation, but the identification and listing of recommendations that will benefit the safety program and address the audit objectives. The team, however, should make reasonable recommendations and suggestions.

Implementation of recommendations and corrective actions is the responsibility of plant management. Plant management should have a program in place to record all recommendations of the audit team, track the progress of plant personnel in implementing appropriate corrective actions, and provide justification for the corrective actions management elects not to implement or to defer to a later date.

ADVANTAGES AND DISADVANTAGES

Safety audits can provide numerous advantages. Findings identify where

- The safety awareness of operating personnel needs to improve.
- Aging equipment or obsolete control systems may become a factor in continued production.
- Deviations or deficiencies exist in operating procedures.
- The introduction of new equipment or the modification of procedures has created new hazards.

It is essential to schedule safety audits regularly, particularly for facilities that contain high risk operations and where plant releases would significantly affect society.

The disadvantage of safety audits is that the results are only qualitative.

For example, an audit may identify a potential hazard resulting from a leak or vessel failure, and thereby identify the volume of liquid that may be released. The procedure will not provide any indication of the frequency with which this may occur.

EXAMPLE AUDITS

Following are two Safety Review Report examples, with appropriate checklists, both of which were prepared to review a chemical facility for potential hazards.

Example: Existing Facility Safety Audit

A detailed safety audit was conducted at a major pharmaceutical production facility. This existing plant has extremely hazardous chemicals on-site in significant quantities. The product is produced in a batch process, but the process is repeated continuously; all equipment is therefore dedicated to the process and maintained accordingly. This particular facility generated updated P&IDs and PFDs to support the safety audit. An extensive operations and emergency response manual is in use and was used for this safety audit. All operators are periodically trained or retrained, and their performance evaluated; records of the training program were available to the audit team for review. The following report was prepared based upon the audit findings.

Example
Safety Audit Report
(Existing Facility)

Date:
Equipment/Process Reviewed: Product Code HES-RT-N and related process equipment
Names of Reviewers: Associated Project Engineer
Supervisor of Maintenance
Chemist
Chemical Engineer
Manufacturing Manager
Scientist
Consultant
Employer of Reviewers: Pharmaceuticals Inc., Anytown, USA
Mr. _____ (Consultant) is an employee of ABC Engineering.

1.a. Visual inspection to determine that the process flow diagrams (PFD), piping and instrumentation diagrams (P&ID), and electrical one-line diagrams (EOL) reflect the actual conditions with respect to the following:

	Yes	No	Comments
Process equipment	x		
Runs and sizes of pipe	x		
Location and function of instruments	x		
Location, function, and size of valves	x		

List below the P&IDs, PFDs, and EOLs reviewed for this stage of the audit
P&IDs: RT.N-PI-001A   PFDs: RT.N-PF-001A EOL: ELEC-MCC-4.1
      MISC-PI-001A   RT.N-PF-001B

1.b. If "no" was entered for any of the above categories, please explain below. Responsible manager must be notified immediately.
NOTE: Attach copies of completed P&ID, PFD, and EOL checklists where appropriate.

2.a. Visual inspection (or review of up-to-date inspection records) to determine that safety devices and emergency systems are consistent with design documents and that these systems are functioning or capable of functioning.

	Yes	No	Comments
Deluges			n/a
Interlocks			n/a
Controls	x		
Backup Systems			n/a
Others: Rupture Disks	x		*

* Rupture disk rating tags checked for consistency with design specifications.

2.b. If "no" was entered for any of the above categories, please explain below.
NOTE: Responsible manager must be notified immediately. Attach copies of completed P&ID, PFD, and EOL checklists where appropriate.

3.a. Review actual (normal) operating conditions of flow, temperature and pressure, process chemistry and raw material feeds against design documents to determine whether the actual conditions are within the limits of the design specifications for each piece of equipment. Are process conditions within the equipment specifications?

List major pieces of equipment (include appropriate tag identification number)

ID Number	Equipment Description	Yes	No	Comment
RT.N-r-26	Glass lined steel reaction vessel	x		
RT.N-r-62	15 Gallon weigh tank	x		
RT.N-r-22	Glass lined steel reaction vessel	x		
RT.N-r-39	Glass lined steel reaction vessel	x		

NOTE: Only major pieces of equipment are listed. Associated equipment (e.g., piping, pumps, instrumentation) was also subject to review during this audit.

4.a. Visual inspection of the facility and process to determine that the actual operating procedures accurately reflect those described in the Operations Manual. This inspection shall include interviews with appropriate operating personnel.

Product Code No: HES-RT-N
OP number and revision: Procedure No. 69, dated March 19 _____
Title of person interviewed:
Chief Chemical Operator
Supervisor, Operations
Manager, Operations
Manager, Inventory and Stores

	Yes	No	Comment
Deviations found and recorded?		x	None Found

| \multicolumn{5}{c}{**SAMPLE P & ID CHECKLIST**} |
| \multicolumn{5}{c}{**(EXISTING FACILITY)**} |

ITEM NO.	DESCRIPTION	YES	NO	COMMENTS
1.	Do the diagrams have adequate legend and citations of referenced documents?	X		
2.	Has all extremely hazardous substance (EHS) equipment (including installed spares) been show?	X		*1
3.	Has all piping been shown, including size, schedule, identification number, ANSI specification and direction of flow?	X		*2
4.	Are appropriate instrument symbols and identification of instruments made, including pertinent instrument functions (e.g., trips, interlocks, etc.) in accordance with the Instrument Society of America (ISA) standards or other satisfactory standards?	X		
5.	Is failsafe position shown for every control valve or non-hand-operated valve?	X		
6.	Are all steam traps, insulation, and heat tracing shown?	X		
7.	Are sizes of important EHS equipment nozzles such as drains, vents, and flushing connections shown schematically to reflect function and elevation?	X		
8.	Are the type, size and set pressure of all relief valves listed?	X		*3
9.	Are all instruments monitoring early detection of abnormal conditions or an early EHS release noted?			N/A
10.	Are the following items noted where critical: a. Relative elevation between equipment and piping?	X		
	b. Slope of piping?	X		
	c. Symmetrical piping?			N/A

Figure 5-1   Sample P&ID checklist (existing facility).

ITEM NO.	DESCRIPTION	YES	NO	COMMENTS
	SAMPLE P& ID CHECKLIST **(EXISTING FACILITY)** **(CONTINUED)**			
11.	Do the notes on P & ID furnish information on the following EHS equipment:			N/A
	Materials of construction?	X		N/A
	Design temperature?	X		N/A
	Design pressure?	X		N/A
	Heat exchanger design thermal duty?	X		N/A
	Rotating equipment design capacity?	X		N/A
	Rotating equipment dynamic head?	X		N/A

NOTES:

*1   Equipment ID numbers were used for this audit.

*2   All items except American National Standards Institute (ANSI) specifications
       are shown; ANSI specifications are not available for this equipment.

*3   This data is shown on the instrument control data.

Figure 5-1   (*Continued*)

Example: New Facility Safety Audit

A detailed safety audit was conducted for a new process facility that intended to use and store ammonia. The ammonia is used to control plant stack emissions by injection into the stack gas. The injection system comprises a storage vessel, evaporators, air compressors, and associated piping and valves. Design basis, P&IDs, (PFDs), and the criteria for design and operation were used to support the safety audit. The plant-specific operations, maintenance, or emergency response procedures had not been developed. To accommodate the safety audit, the operations, mainte-

	SAMPLE PROCESS FLOW DIAGRAM (PFD) CHECKLIST (EXISTING FACILITY)			
ITEM NO.	DESCRIPTION	YES	NO	COMMENTS
1.	Does diagram depict the use, generation, storage, and handling of the EHS raw materials and product?	X		*1
2.	Is the flow of the material from item to item clearly marked on the diagram?	X		
3.	Are the basic control loops or major control schemes marked clearly on the diagram?		X	*2
4.	Are points of discharge to the air, sewer, and waste collection receptacles clearly indicated on the diagram?	X		
5.	Are any other potential points of discharge to the environment, such as rupture disks, indicated on the diagram?		X	*2
6.	Does the diagram have an adequate legend?	X		
7.	Are appropriate cross-references (where applic-able) shown on the diagram which refer to documents and give details of material balance, flows, raw materials, products, intermediates, treatment chemicals, operating conditions (temperature, pressure, batch or continuous)?	X		

NOTES:

*1   Handling and use of the extremely hazardous substance is shown in the PFD AC.B-PF-001A. Storage of the extremely hazardous substance is addressed in a separate safety audit.

*2   These items are shown on P & ID AC.B-PI-001A.

Figure 5-2   Sample process flow diagram (PFD) checklist (existing facility).

nance, and emergency response plans for a similar facility were employed. Operator training had not yet been established and was not considered during this audit. The following report was prepared based upon the team's findings.

SAMPLE ELECTRICAL ONE-LINE DIAGRAM (EOL) CHECKLIST (EXISTING FACILITY)				
ITEM NO.	DESCRIPTION	YES	NO	COMMENTS
1.	Are all end users (i.e., power consumers) shown on the EOL including consumption requirements, e.g., motor xyz, 50hp, 60 hz, AC?	X		
2.	Are connections to motor starters, or control centers, and all intermediate equipment shown?	X		
3.	Are connections to power distribution centers shown including transfer switches and backup tie feeders?	X		
4.	Are interconnections from substations to main feed shown?	X		
5.	Are emergency power sources (EPS) shown?	X		
	Identification of equipment load shed when EHS is on-line?	X		
	Rating for each EPS?	X		
6.	Are the operating statuses (e.g., normally open or closed) depicted for all circuit breakers or switches associated with main feeders or services?	X		Normally Closed
	END OF EXAMPLE ONE			

Figure 5-3   Sample electrical one-line diagram (EOL) checklist (existing facility).

SAMPLE
AUDIT REPORT
(NEW FACILITY)

1. The following documents were reviewed to ensure that they are consistent, reflect the state of the art and criteria in design and anticipated operation.

DeNox System Description   Doc. No. ZBCKR/22*22

DRAWINGS

P&ID Thermal DeNox System   Dwg. No. M27, Rev. 3
P&ID Combustion Air and Flue   Dwg. No. M13, Rev. 7
  Gas System
Heat and Mass Balance   Dwg. No. M-5, Rev. 5
4160V Single-Line Diagram   Dwg. No. E-2, Rev. 3
Motor Control Center (MCC)   Dwg. No. E-4, Rev. 3
  Single-Line Diagram
General Arrangement Plan E1.
  (−)25′-0″, 0′-0″, & 15′-0″   Dwg. No. M-2, Rev. 10
Electrical Area Classification Plan   Dwg. No. M-300, Rev. 2
Yard Piping Plan   Dwg. No. M-26, Rev. 14
Plot Plan Overlay on Sketch Plot   Dwg. No. C-6, Rev. N
Sketch Plot (By XYZ Associates)   Dwg. Dated Aug. 1987

SKETCHES

Process Flow Diagram
  Anhydrous Ammonia
  DeNox System   __/__/__   SK-M-200, Rev. A
Anhydrous Ammonia Storage
  Tank   __/__/__   SK-140-AIS-TK, Rev. 1

SPECIFICATIONS

Spec. Sheets for Anhydrous Ammonia Storage Tank   __/__/__
Spec. Sheet for Anhydrous Ammonia Vaporizer   __/__/__
Spec. Sheets for Carrier Air Compressor   __/__/__
Spec. Sheets for Anhydrous Ammonia System Excess
  Flow Valves   __/__/__

Spring-Flex Type MFP Metal Flexible Hoses Style
  MFE (Fixed Flanges)   *
Spec. Sheets for Control Valves, Flow Transmitters,
  and Pressure Transmitter   ___/___/___
Spec. Sheets for Compressed Air Piping (Class 150-1) ___/___/___
Spec. Sheets for Ammonia Piping (Class 300-1)   ___/___/___
Thermal DeNox System Equipment List   ___/___/___

* Undated; Used copy hand-dated ___/___/___ by Mr. Human

Note: The basis of the state-of-the-art reviews was U.S. Dept. of Commerce NTIS, Prevention Ref. Manual Vol. 2 Control of Accidental Release of Ammonia, Aug. 87 (Doc. No. PB87-231262)[1]

2. The following components were checked to ensure that they reflect design parameters (i.e., temperature, pressure, flow) consistent with system design criteria:

Ammonia Storage Tank (AIS-TK-1)
Ammonia Vaporizers (AXS-VAP-1A to 1D)
Boilers (Nos. 1 to 3)
Ammonia Piping and Valves

3. The documents listed in Item 1 for the following safety devices were checked to ensure that the devices will perform their intended function:

Relief Valves for Equipment and Lines Containing Ammonia
Ammonia Storage Tank (PSV-1&2)
Ammonia Vaporizer (PSV-882 to 885)

Controls and Interlocks
Ammonia Vaporizer Temperature, Pressure and Level
Ammonia Vapor to Boilers, Pressure and Flow
Boiler Air/Ammonia Flow
Ammonia Analyzer & Alarm (AE-843)

4. The following deviations were brought to the attention of the responsible manager (see Note) on April 5, 19 __:
   a. At the time the Safety Review was conducted (April 5, 19 __), the information upon which the design was based had not been properly

consolidated into the Criteria for Design and Operations (CDO). However, after reviewing the current, final CDO (during the week of May 22), it is noted that the basis of the design has not substantially changed.

b. A thermal relief valve discharging to the diked area slab should be installed in the liquid ammonia line from the tank to the vaporizers unless calculations are prepared and they indicate that entrained liquid pressure will not exceed line design pressure under postulated temperature rise.

c. Provide protective hoods to preclude water accumulation in all ammonia vapor relief valve discharge stacks. (See Butala[1] for further details.)

d. Add high-level alarm to ammonia storage tank to warn of impending overfill.

e. Consider the installation of a scrubber for vapor discharge from ammonia storage tank and vaporizer relief valves to minimize or eliminate ammonia release to atmosphere.

f. The installation of a shed over the ammonia storage tank and dike area should be considered. This will inhibit tank overheating due to exposure to direct sunlight and will also help vapor containment in case of tank overfill or rupture.

g. Add strainer upstream of self-contained regulator PCV-123a to protect it from pipe construction material and dirt.

h. Consider provisions for purging vapor space of ammonia storage tank on a regular basis with an oxygen-free inert gas to reduce possibility of stress corrosion cracking.

i. The data sheets for the ammonia storage tank and the vaporizers should state that the relief valves are to be constructed in accordance with Compressed Gas Association (CGA) S.1.3, Part 3.

j. Pressure regulator PCV-456 should be flanged for easy removal. Although Butala[1] also recommends that solenoid valves be flanged, this installation incorporates redundant equipment installations so that ease of removal is not critical to continued plant operation.

k. Route ammonia truck vapor equalizing line and, if possible, liquid line over dike rather than through it.

l. Make provision to remove rainwater and ammonia from ammonia truck pad sump. Suggest using portable pump indicated for same service for diked area.

m. Consider making provisions for waterspray barriers to mitigate ammonia vapor release.

n. The ammonia storage tank, data sheet item 4i, above, should be revised to require 100 percent stress relieving after fabrication to reduce the risk of stress corrosion cracking.

o. Provide sloped lines for draining condensate from ammonia vapor relief valve discharge.

p. Consider making instruments redundant in order to improve system availability.

NOTE: Mr. D. E. Hacks, Q Services, Inc. Process Engineer, was notified of the deviations on the date indicated. He will be responsible for informing the responsible manager.

5. The following inconsistencies were identified between documents listed in Item 1:

a. The DeNox System Description on page 6 should note that valves will be installed under all pressure switches and gauges. These valves are not shown on the P&ID for the vaporizer pressure gauges.

b. The P&ID indicates that each vaporizer is furnished with high-level switches. These are not indicated on the equipment data sheet or noted in the system description.

c. The tank HI-HI/LO-LO pressure switches noted in the System Description are shown on the vaporizer vapor line to the tank. The System Description states that these switches are furnished on the ammonia storage tank itself.

d. The Vaporizer Data Sheet notes a dual element thermocouple for operational control for high-temperature shutdown safety. Only a single temperature element is indicated in the P&ID.

e. The System Description notes that the feed valve installed at the bottom of the vaporizer shell will be used for sampling and purging. The vaporizer data sheet notes this to be a drain only.

f. The Ammonia Piping Spec. Data Sheet, Item Ba, states in part ". . . flexible connections are not acceptable for ammonia transport lines." Item Bc states, "All metal flexible connections for permanent installations shall have a minimum working pressure of 250 psig and shall be stainless steel."

g. Yard Piping Plan Dwg. MA-26 Rev. 14 shows guardrail and curb around Ammonia Storage Tank and Ammonia Unloading Station. General Arrangement Plan Dwg. MM-2 does not show guardrail or curbs.

h. From Ammonia Storage Tank Dwg. SK-140-AIS-TK-1, it appears that nozzle No. 11, rotary gauge size 1", is unused, since no connection is shown on the P&ID.

i. Under Calculation of Tank Size Requirements, NH_3 feed is given as 335 lb/hr at 153.3 psig, 80°F. Heat & Mass Balance Dwg. MM-5 shows 404 lb/hr at 153.3 psia, 80°F.

j. The System Description does not contain specific information on the acceptable range of ammonia usage, reactions, products and by-products.

6. Corrective actions taken pursuant to deviations found in Items 4 and 5 or justification for deviation from state of the art are as follows:

Item 4a. A single comprehensive document has been prepared containing the criteria for design and operation of the EHS system.

Item 4b. A thermal relief valve has been added to the liquid ammonia line.

Item 4c. Protective hoods have been added to the relief valve stacks.

Item 4d. A high-level switch and alarm have been added to the ammonia storage tank.

Item 4e. A scrubber is not being installed due to the release rates associated with the safety relief valve release scenarios (i.e., [1] solar heat input = 0 (zero), and [2] vessel exposed to fire < 5RQ). The vessel will be protected against fire exposure by

1. Remote location relative to normal on-site traffic.
2. Protection against vehicular crash by dike and roadside rails/posts.
3. Location of permanent fog hose stations at fire hydrants near storage location which will be used to cool the tank and fight a fire near the tank. Fog nozzles also can be used to mitigate vapor releases due to spills.

Item 4f. A shed is not considered necessary, because of the regional location of the plant in the Northeast. Calculations indicate that solar load cannot raise the vapor pressure of the ammonia in the tank to the 250 psig relief valve set point so that no vapor releases are predicted due to solar heating. In addition, the tank will be painted white to minimize solar absorption. The tank will be provided with a local high-level alarm to warn the delivery operation of an overfill condition. The vapor return line has been provided with a high-pressure switch, which alarms in the main control room if the tank pressure exceeds the alarm set point. Initiation of this alarm would require emergency operator response.

Item 4g.   The design has been revised. PCV-123a pressure control valve and FCV-456 flow control valve have been replaced with a single valve actuated by a pressure controller and interlocked to close on high ammonia flow. A strainer is therefore not required.

Item 4h.   Before the ammonia storage tank is filled with ammonia, it will be purged with nitrogen.

Item 4i.   The data sheet for the ammonia storage tank has been revised to require that the tank relief valves be constructed in accordance with CGA S.1.3, Part 3.

Item 4j.   The design has been revised. See Item 4g.

Item 4k.   The final design will route the ammonia truck vapor equalizing line and the liquid ammonia truck fill line over the dike.

Item 4l.   The system description has been revised to note that a portable pump will be used to empty the ammonia truck pad sump. This will also be incorporated into the operating procedures.

Item 4m.   The two fire hydrants in close proximity to the ammonia storage tank dike have been equipped with hose stations and fog nozzles. Their use in controlling ammonia vapor release will be incorporated into the plant emergency procedures.

Item 4n.   The ammonia storage tank data sheet has been revised to require 100 percent stress relieving after fabrication.

Item 4o.   Data sheets for ammonia vapor relief valves have been revised to require a drain hole in the valve outlet to facilitate removal of relief stack condensate buildup.

Item 4p.   Instrument redundancy is not considered necessary to improve system availability. Instruments can be isolated and replaced in the existing design.

Item 5a.   The P&ID has been revised to include isolation valves for the vaporizer pressure gauges.

Item 5b.   The vaporizer data sheet has been revised to include the high-level switch.

Item 5c.   The System Description has been revised to note that the HI-HI/LO-LO ammonia vapor pressure switches are on the common vapor line from the vaporizers to the storage tank.

Item 5d.   The Vaporizer Data Sheet has been revised to note that one element of the dual element thermocouples is a spare.

Item 5e.   The System Description Vaporizer Data Sheet and P&ID have been revised to indicate that the valved nozzle at the bottom of the vaporizers is a test and drain connection.

Item 5f.   The System Description and Ammonia Storage Tank Data Sheet have been revised to note that the gross capacity is 18,000 gallons nominal and the useable content is 87 percent of capacity, or 15,660 gallons.

Item 5g.   The Ammonia Piping Spec. Data Sheet has been revised to allow the use of metal flexible hose.

Item 5h.   Drawing MM-2 has been revised to show the guardrail.

Item 5i.   The Ammonia Storage Tank Drawing has been revised to indicate that nozzle No. 10a is to be used for a high-level switch. (See Item 4d.)

Item 5j.   The ammonia storage tank capacity calculation has been revised to indicate a usage rate of 404 lb/hr.

Item 5k.   A table indicating the range of ammonia usage, reaction products, and by-products has been added to the System Description.

SUGGESTED READINGS

1. Butala, D. M. September 1985. Safety audit. *Chemical Age India* 36(9).
2. Atkins, D. May 1978. Safety audit. *Chart Mechanical Engineering* 25(5).
3. Brown, W. J. 1981. *Safety management of complex research operations*. Paper read at Fifth International System Safety Conference. Denver, CO.

6

WHAT-IF Analysis

William W. Doerr

The WHAT-IF methodology is one of the three simplest forms of conducting a hazard analysis; the other methods are checklists and the preliminary hazard analysis discussed in Chapters 3 and 4, respectively. Unlike other, more complicated methods, such as Dow/Mond analysis, fault tree analysis, and hazard and operability studies (HAZOP), this technique does not require special quantitative methods or extensive preplanning. The method does utilize input information specific to a process to generate a set of checklist-type questions. A special team prepares this comprehensive list of questions, called WHAT-IF questions, which are then collectively answered by the team and summarized in tabular form[1].

The WHAT-IF technique is widely used during the design stages of process development, as well as during the operating lifetime of a facility when improvements or changes are being made to the process or operating procedures. In the latter case, deficiencies need to be identified before a change is implemented, and the intent of the method is hazard identification and prevention. Generally, the hazards identified are acute or accidental hazards, not chronic exposure hazards.

DESCRIPTION OF METHODOLOGY

The methodology behind a WHAT-IF analysis is a speculative process whereby questions in the form "What if . . . ?" are formulated and reviewed. The method has the following basic features: scope definition, team selection, review of documentation, question formulation, response evaluation with consequences, summary tabulation to the set of questions.

The basic characteristics of a WHAT-IF analysis and the features marking a thorough analysis are as follows:[2]

- A WHAT-IF analysis should be a systematic study of:
  updated process chemistry
  operating procedures

maintenance procedures
operator job descriptions
process flow diagrams
piping and instrumentation diagrams
on-site chemical inventories
other design documents for the facility

- A WHAT-IF analysis should be performed by a multidisciplinary team.
- The list of WHAT-If questions should be prepared in advance.
- The question list may be prepared in conference or independently.
- The WHAT-IF team should be formed so that it is able to answer questions with minimal outside assistance.
- The team leader and the recorder are assigned for the duration of the study.
- The analysis results must be presented in a tabular format and include:
WHAT-IF questions
description of corresponding consequence or hazard
an assessment of criticality based on the potential release rate.

Scope Definition

With this method it is important to first define the type of process hazards that limit the scope of the study. These hazards may include any of the following: fire, explosion, gaseous toxic chemical release, liquid toxic chemical release, radioactive release, improper reaction, or odor release. The next step is to define the physical boundaries of the source of the potential hazard and the impact area or receptor resulting from the potential hazard. These physical boundaries are defined by the team leader or initiator of the study.

Source and Boundaries

The physical boundaries for the source of a process hazard may be (1) a single piece of equipment, (2) a process unit with several pieces of equipment, (3) an entire facility or collection of process units, or (4) a community. Definition of the source of the process hazard is important, because it defines both the extent of a *source term* (an industry term meaning "location and amount of hazardous material released"), as well as the interfacing equipment considerations. An example of the latter is utilities such as steam, electricity, process control, or instrument air that cross the process boundary because they are supplied elsewhere. It is useful for the WHAT-IF analysis to know what happens if an interfacing utility is lost. The consequence may be that some equipment may fail in a safe manner, whereas other equipment may fail in an unsafe manner, for example, because of a loss of instrument air. A tabular listing of interfacing items should be prepared and reviewed for the potential impact they would have on the subject process when they deviate from their normal operating ranges during question preparations for the WHAT-IF analysis.

A community may be considered a physical boundary for the source of a process hazard if transportation considerations are important for materials delivery and exit from the operating facility. Examples are a 90-ton rail car of chlorine, a liquefied petroleum gas (LPG) pipeline to a facility, or gas cylinders of any toxic chemical shipped to a site through the surrounding community. While hazards associated with the use of these compounds at the site may be important, it may be equally important to address occurrences outside the boundaries of the site during shipments to or from the site. On-site hazards analyses are by definition limited to the site boundary. Analysis of hazards associated with transportation of hazardous substances are usually the responsibility of the transporter; a chemical transfer accident is usually the responsibility of the site owner. This is an important distinction, because transporters must conform to a distinct set of standards (e.g., U.S. DOT).

The definition of the impact area or receptor area for a WHAT-IF analysis is required. The consequences recognized by a WHAT-IF analysis may be categorized into three general groups: (1) consequences that affect workers on the site, (2) consequences through which economic loss or production downtime occurs and, (3) consequences through which the off-site population is affected. The physical definition of the process hazard and the physical area affected by the hazard are needed later in the study when the hazard is summarized and ranked with respect to potential effect.

Team Selection

After the scope of the study is defined, the second step in the WHAT-IF analysis is the selection of the team members who will perform the study. The selection should result in a multidisciplinary team that can contribute in various subject areas such as operations, engineering design, maintenance procedures and practices, and safety or plant emergency practices.

The team should therefore include at least one of each of the following: a process operator familiar with startup, shutdown, and normal and abnormal operating situations; a process engineer familiar with the design, function, and safety features of the process; a maintenance engineer familiar with the inspection and maintenance practices for the subject equipment; and a safety or plant manager familiar with overall on- and off-site issues.

The team is led by the team chairperson or leader who (1) defines the scope as described above, (2) organizes the team members, (3) organizes the questions and their associated review by the team, and (4) selects a clerk who summarizes the responses to the questions during the review. The team leader is important during the review process because the leader maintains the focus on the previously defined scope and physical boundaries.

Documents

The next and third step of the WHAT-IF analysis is the selection and review of documents necessary for the preparation of WHAT-IF questions. These documents are discussed in greater detail in the section entitled "Supporting Documentation." These basic documents should be reviewed by each team member individually before the WHAT-IF review and analysis are performed by the team. During the individual document review, questions may be formulated that pose hypothetical upsets to, or deviations from, the normal operation of the process.

Question Formulation

The formulation of WHAT-IF questions by individual team members represents the fourth step in the WHAT-IF analysis. Questions are formulated in the style of the question, "WHAT IF . . . ?". The questions may address any of the following:

Equipment failure
Process condition upsets (due to temperature, pressure, or feed upsets)
Instrumentation failure
Interfacing utility failures
Operator improvisation, malperformance, or inattentiveness
Departures from operating procedures during normal operations, startup, or shutdown
Maintenance-related accidents
Site-related features, such as area transportation, impact with a crane, or handling-related accidents
External events, such as airplane accidents, vandalism, or storms, and
Combinations of failures, such as multiple equipment failures, or equipment failure combined with operator error.

Questions are best prepared in a systematic manner, by starting at the feed or start of a process and proceeding through the process to its completion or outlet product stream. This chronological evaluation can also be applied (although it is not often done) to an operating procedure, where upset conditions may be evaluated with the steps of an operation.

Sometimes, the formulation of the WHAT-IF questions is deferred until the fifth step, the response and evaluation from the WHAT-IF team review. When the questions are deferred, the team leader should review the process or operation in a chronological manner, so that questions are methodically prepared, answered, and recorded with the group response to the question.

WHAT-IF Evaluation

Once the questions are prepared by each member of the WHAT-IF team, the team leader proceeds to the fifth step of the WHAT-IF analysis. The questions are collected by the team leader, who consolidates the questions into a collective set of questions, which should be arranged by the chronology of the process or operation. These questions should be listed as shown in Table 6.1, with the column headings usually including

- Question description
- The corresponding consequence or hazard
- The criticality of the hazard based upon the potential release
- The recommended actions to mitigate the hazard.

The table is incomplete before the team review begins; each row in the table represents an individual WHAT-IF question. The associated consequence, criticality, and recommended actions are determined by the WHAT-IF team during the team review. The team review is now orchestrated by the team leader, who formulates the team's response to each category for every question. It is usually a good idea for the team leader to use an overhead transparency of the process being discussed, so that questions regarding the process can be quickly resolved with the aid of the drawing (which may be a process flow diagram, or a piping and instrumentation diagram). All answers to the various categories are obtained from the team by the team leader before the next question is reviewed. Various considerations for the responses are discussed further in the "Results" section of this chapter. The record-keeping (i.e., responses to each question) is best performed by the team clerk, and not by the team leader, so that the discussion can proceed quickly and effectively.

The WHAT-IF team should be able to answer each question without any outside assistance. While this may not always be possible, the team mem-

Table 6-1 WHAT-IF Hazard Analysis Table

	WHAT-IF Question	Answer/Hazard	Criticality	Possible Recommendation
18.[a]	The FCC main column overhead vapor line ruptures? Note: This applies to all lines connected to main column and main column overhead receiver.	Emergency shutdown procedure would be initiated; inventory of reactor and column and gas concentration unit (until it is blocked in) would be released to the atmosphere. Since this stream is cooler than the reactor outlet, potential for autoignition is reduced, so H_2S may not be combusted.	M[b]	Follow standard procedures and practices. Notify state and local responsible agencies.
23.	The level gauge on the main column overhead receiver breaks?	H_2S in a hydrocarbon vapor stream would be released until operator blocks in level gauge. Operator would be alerted by erratic pressure and level readings in control room and level alarms.	M	Follow standard procedures and practices.
28.	Tube in interstage cooler ruptures on gas concentration unit (GCU)?	Hydrocarbon would be released into cooling water. The H_2S would be stripped by the air exiting the cooling tower.	M	Hydrostatically test bundles and plug leaks when mechanically cleaned.
31.	Tubes in high pressure cooler ruptures on GCU?	Same as answer to 28.	M	Same recommendation as WHAT-IF question 28.
38.	Tubes in primary absorber lower intercooler ruptures on GCU?	Same as answer to 28.	M	Same recommendation as WHAT-IF question 28.
39.	Tubes in primary absorber upper intercooler ruptures on GCU?	Same as answer to 28.	M	Same recommendation as WHAT-IF question 28.
41.	Tubes in lean oil cooler ruptures?	Same as answer to 28.	M	Same recommendation as WHAT-IF question 28.

53.	Tubes in dubanizer condenser ruptures or header leaks?	Vapor containing H$_2$S would be released to the air. Operator inspects area weekly. Hydrogen sulfide monitors in area would alert operators of release. Bundle would be removed from service and leak repaired.	M	Follow standard operating procedures and practices.
54.	Tubes in liquefied petroleum gas (LPG) cooler leaks?	Same as answer to 28.	M	Same recommendation as WHAT-IF question 28.
56.	Overhead flange on sour water stripper leaks?	Stop feed to stripper and steam out. Sour water tank normally operates with 10 feet level and 30 feet is available freeboard. Monitors are located in this area and operator can take prompt action as a result of monitor alarm.	M	Follow normal shutdown procedures and inspect piping periodically.
59.	The feed thermal relief malfunctions?	Noticed by field operators and blocked off until it is repaired. Negligible release.	M	Inspect relief valve periodically.
61.	High Reid vapor pressure (RVP) hydrocarbon is routed to sour water tank?	Tank is provided with a double seal floating roof that limits the amount of H$_2$S released to the atmosphere. Operations personnel locate the malfunctioned level controller and notify state.	M	Review the possibility of installing H$_2$S monitor and alarm. Initiate emergency notification procedure.
63.	Monoethanol amine (MEA) regenerator overhead vapor line ruptures (external event)?	Refining facility shutdown. Release of H$_2$S until the rich MEA flow to the regenerator is stopped. Detected by loss of flow/pressure on control room indicators. Hydrogen sulfide monitor in the area detects release	M	Initiate emergency response. Follow normal shutdown procedures. Piping inspection program is followed while unit is on line.

(*continued*)

Table 6-1 WHAT-IF Hazard Analysis Table (Continued)

WHAT-IF Question	Answer/Hazard	Criticality	Possible Recommendation
	and alarm, alerting personnel in the area and control room.		
70. Level gauge on regenerator reflux drum at fuel gas treating breaks?	Incorrect level on the reflux drum. H_2S monitor alarm sounds and operator isolates leak. Minor release.	M	
73. Drain valves on knockout drums are left open?	H_2S to sewers. Detected by monitors in the area. Minor H_2S release. Control room operators are alerted by alarms in plant control system.	M	Field operators check level at regular intervals. Low point drain valves normally blinded.
80. Controller is relieving to flare and the flare pilots are out?	The plant will be shut down. H_2S is released for about 30 minutes.	M	Follow normal shutdown procedures and emergency notification procedure.

[a] *Not all of the questions appear because only moderate criticality hazards are represented.*
[b] *M = moderate. Other values of criticality are "low" and "high."*

ber selection process should help ensure a high probability of providing answers to all questions from team members. Should a WHAT-IF question be unanswerable by the team, the team leader should note this on the table and defer to the appropriate individual not in attendance. The team leader should formulate an answer for the summary table after consultation with the nonmember.

The sixth and last step in the WHAT-IF analysis is the completion of the summary table prepared by the team review. The table should include not only the four column headings used in the team review, but also traceable information, such as unit identification, location, date, drawing number, revision number of drawing, and team participants. The table should also reflect any follow-up items discussed during the team review that were unresolvable at that time.

SUPPORTING DOCUMENTATION

Documents important for the WHAT-IF analysis and the preparation of the WHAT-IF questions include at least the following items:

- *Process chemistry description*. This document is important for complicated processes in which chemical reactions or phase changes occur. Usually this document provides information about the raw and product materials and their purity, as well as information on any intermediate or side reactions. Special chemical problems are sometimes discussed, such as runaway reactions, handling problems (such as explosive dusts or cryogenic materials), or unusual physical properties.
- *Operating procedures*. These documents frequently include procedures and conditions for normal operations; definition of and control procedures for abnormal conditions; definition, control, and mitigating procedures for emergency conditions; pre-startup and startup procedures; shutdown procedures; discussion of safety devices; material safety data sheets; and any checklists for these operating procedures.
- *Maintenance procedures*. These documents are useful for identification of equipment that has preventive maintenance schedules and inspections; work order controls or permits for equipment modification; testing or rebuilding requirements for certain equipment; functional checking of certain equipment; testing of standby emergency equipment; and checklists for confined space entry, lockouts, and hot work.
- *Operator job descriptions*. These documents help to identify the duties and responsibilities for each operator, usually including data collec-

tion; use of checklists; equipment startup and shutdown; normal equipment operation; familiarization with material safety data sheets; and emergency procedures.

- *Process flow diagrams (PFDs).* These diagrams show or define the flow connections (piping) between the various vessels and other equipment in the processes; the feed streams, the product and waste streams, the stream compositions and properties (such as temperature, pressure, and phase); operating cycles and batches, and sometimes equipment sizing (capacity, area, throughput, etc.); and material balances. Such information is needed to define the amount of material (i.e., source term) that potentially may be released.

- *Piping and instrumentation diagrams (P&IDs).* These drawings provide additional information that supplements the PFDs. This information includes identification of every pipe (with size, flow direction, and piping specifications); identification of every control loop and instrument function; indication of all valves with their fail-safe positon, if appropriate; data on set-point conditions for safety relief devices; and references to other interfacing drawings. These drawings are useful to show additional information on connections not shown in PFDs, set points, and other interfacing equipment.

- *Hazardous material inventories.* These documents for hazardous substances may already be prepared and available as required by the Resource Conservation and Recovery Act (RCRA), the Superfund Amendment and Reauthorization Act (SARA) Title III, the EPA Form R, The New Jersey Toxic Catastrophe Prevention Act (TCPA), OSHA CPL-2.2, or the California Risk Management Prevention Plan. These documents are important to reveal any oversights not shown on the above documents.

- *Other documents.* These documents may include site plans, equipment layout drawings, emergency response plans, external forces and events summary for weather or other related external hazards, or electrical area classification drawings for the site.

In addition to the documentation, it is important for the team leader to perform a site tour of the hazardous operation as part of the WHAT-IF analysis. This allows the team leader to become familiar with the general condition and interconnection of the equipment at the site.

It is beneficial to have a site plan and a process flow diagram in hand during the site tour, in order to review and compare the equipment observed in the field with the drawings. This comparison may give rise to additional questions, which the team leader can prepare and include in the evaluation.

METHODOLOGY IMPLEMENTATION

The WHAT-IF analysis is beneficial both during the design phase of a new project and during a design or operating procedures change for an existing process. The method considers the effects of unexpected occurrences that would yield an undesirable result. In this way, possible undesirable occurrences may be defined before a process is fully designed or before a design change is implemented. Feedback from the WHAT-IF analysis on a new or modified process is beneficial, because changes to the process can be made before an operation is implemented and before any hazardous operation is fully realized.

The WHAT-IF analysis requires the selection of a team leader, who, in turn, selects the team members. The criteria for the selection of the team members have been previously discussed. The team leader's goal is to organize and focus the group during the course of the study. The selection of appropriate team members is critical, because they can provide the imagination and broader process or site perspective that the team leader may not have or be qualified to evaluate.

During the group review meeting, it is important for the team leader to provide continuing motivation to the group, moving them forward in responding to questions, rather than allowing them to become preoccupied or absorbed with the answer to a single question. If an answer is unknown, it should be deferred until after the session if a specialist is required. As a rule, it is unwise to spend more than a few minutes per question. Like any group evaluation, the WHAT-IF analysis should not last longer than a few hours per session. If more than one session is required, the session should be broken up so that each is less than a few hours in duration.

RESULTS

The results of a WHAT-IF analysis are summarized in a table such as the blank form in Figure 6-1. This is a sample table, and local regulatory requirements may necessitate alteration of its appearance or contents. The WHAT-IF question appears in the first column. The hazard associated with the question is found in the second column. The criticality ranking is found in the third, and the action for further follow-up is summarized in the fourth. Each row in Figure 6-1 represents one WHAT-IF question and the issues associated exclusively with that question.

The criticality ranking is used to indicate the relative qualitative ranking of the hazard associated with the WHAT-IF question. A criticality ranking system should be established by the team leader during the scope-definition phase of the study, before the actual study is undertaken by the

UNIT: _____ DATE: _____ P & ID NO.: _____ REV. # _____

LOCATION: _____ TEAM EVALUATORS: _____

"WHAT IF" QUESTION	HAZARD OR CONSEQUENCE	CRITICALITY RANKING	ACTION FOR FOLLOW-UP
1.			
2.			
3.			
4.			
5.			

Figure 6-1   Sample form for results of WHAT-IF analysis.

team. Some criteria that may be used for the criticality ranking are as follows:

- Estimated release amount. This amount may be expressed in qualitative terms, such as *none, low, medium,* and *high;* the terms relate to some fractional amount of a threshold value, such as a reportable quantity (for RCRA/CERCLA), or a registration quantity (for New Jersey TCPA).
- Number of affected workers on site. This number may be expressed as *all* (general evacuation), *major* (approximately 50% of employees affected, or partial evacuation or sheltered response), *local* ($\leq$ 10% of employees affected), or *none*.
- A release that requires reporting to an agency. This includes a regulatory response, full in-house investigation, management summary, or administrative follow-up.
- Equipment loss. This may be expressed in terms of dollar value (*none, low, medium,* or *high*), using whatever intervals are appropriate for such value of downtime (*none, minutes, hours, days,* or *months*).

The criticality ranking is subsequently used with the summary table to rank-order and establish a priority for the findings of the WHAT-IF analysis for subsequent follow-up. The number of levels within the ranking is arbitrary, since the evaluation is qualitative; however, if certain threshold values are required for the hazard review to initiate regulatory work for compliance, then the number of levels in the ranking should reflect these criteria as noted previously.

For example, New Jersey's TCPA regulations require a quantitative risk assessment for releases greater than 5 times the registration quantity (RQ). Therefore, a typical criticality ranking in terms of estimated release amounts could include *none, low* (defined as less than one RQ), *medium* (defined as between one RQ and five times RQ), and *high* (defined as greater than five times RQ).

At this point, the hazards analysis team may prepare recommendations, since a detailed evaluation will usually be performed to determine an appropriate risk reduction measure or remedial action. The recommended "action for follow-up" in Figure 6-1 provides a response to be taken by the team or its management to reduce either one of the following: (1) the amount of material released by the hazard, (2) the probability of the initiating event associated with the hazard, (3) the amount dispersed by the environment, or (4) the population affected by the hazard. These remedial, mitigative, or preventive actions become the basis for a risk reduction plan, where higher priority for action would be applied to those hazards with a higher criticality ranking in the WHAT-IF analysis.

ADVANTAGES AND DISADVANTAGES

The advantages of the WHAT-IF analysis are as follows:

- Its simplicity (i.e., no specialized technique or computational tool is required).
- Its usefulness any time during the life of a plant (either during the design stage or during a plant modification).
- Its relatively low cost (minimal preplanning is required and no special tools are required; the only major cost is the time associated with the personnel conducting the review).
- The simple tabular summary of the hazard and its potential consequence.

The disadvantages of the WHAT-IF analysis are as follows:

- The requirement for a team of people to perform the study. The team leader may have difficulty in scheduling the various team members to conduct the study, because the staff from the operating group will have other commitments and may be unable to allocate several hours per day to perform the study over an extended period of time.
- The reliance on the experience and intuition of the team members both to develop imaginatively and to solve the WHAT-IF questions.
- The subjective nature of the analysis. The process is subjective because this less analytical method relies on imagination to create and solve WHAT-IF questions; the failure modes and effects analysis (FMEA) or the HAZOP discussed in Chapters 7 and 8, respectively, are less subjective methods.
- The qualitative results obtained. The criticality ranking gives only a general level relative to the objective of the study (i.e., low, medium, or high); a probabilistic risk assessment is quantitative by determining the probability of the event leading to the hazard, as well as a better estimate of the amount released by the hazard.

These advantages and disadvantages are summarized in the following list:
Advantages

- No specialized technique required.

• Can be done any time during plant life.
• Fairly inexpensive.
• Tabular summary of results.

Disadvantages

• Requires a team of people.
• Relies heavily on team experience, intuition, and imagination.
• Subjective, not as systematic as some others (e.g., HAZOP).
• Qualitative results, but with no numerical prioritization.

The decision to select the WHAT-IF method over another method should be based on the level of detail desired in the analysis, the number of persons to be committed to the study, and whether the study is to be an initial scoping study that may later be followed by a more quantitative analysis. If moderate labor hours and a qualitative assessment of the criticality ranking are desired, then WHAT-IF analysis may be preferable over other methods such as a HAZOP or an FMEA. Under these circumstances, the WHAT-IF study is preferable to the hazard assessment methods, such as fault tree or event tree analysis, because the latter methods are often very quantitative. If minimal personnel are available to perform the analysis, and a simpler assessment of the hazards is desirable, then a checklist approach to evaluating the hazards may be preferable over the WHAT-IF analysis.

EXAMPLE WHAT-IF ANALYSIS

A hazard analysis for a refining facility that handles hydrogen sulfide was performed using the WHAT-IF method. The hazard analysis team consisted of refinery personnel (a process engineer, the safety manager, the operations manager, and the refinery manager) and an outside facilitator. The WHAT-IF questions were formulated by the team before the start of the analysis. Equipment failure data, accident reports, and external events were also considered in the preparation of the questions.

This refining facility had recently been refurbished and recommissioned; therefore, the design of each unit utilized current commercially available technology. The WHAT-IF questions presented in Table 6-1 represent those hazard analysis scenarios in which a potential for a hydrogen sulfide release existed. Equipment containing small quantities of hydrogen sulfide had been designated as hazardous equipment by the refinery but was not included in the hazard analysis, since the refinery processed less than

40 lb/hr of hydrogen sulfide, a quantity small enough not to cause any problems beyond the site boundary.

The results of the WHAT-IF analysis in Table 6-1 are presented as examples of the type of information that may be presented in a summary table. The criticality ranking of hazards presented in the table are "M," or "moderate." The recommendations summarized in the table are examples of emergency response or operator follow-up; attention to detail in administrative controls to minimize risk is required where no further actions may be taken to reduce risk. In these situations, a knowledge of emergency procedures is required.

REFERENCES

1. Center for Chemical Plant Safety. 1987. *Guidelines for Hazard Evaluation Procedures*. New York: American Institute of Chemical Engineers.
2. American Chemical Society. 1985. *Chemical Process Hazard Review*. ACS Symposium Series 274. Washington, DC: ACS.

7

Failure Modes and Effects Analysis

Robert L. O'Mara

A failure modes and effects analysis (FMEA) is an examination of individual components made to assess the effect of their failure on subsystems and systems with an emphasis on hardware failure.

The FMEA is a qualitative inductive method that is straightforward and easy to apply. It is typically documented in a tabular format in which the table column headings show its progressive development. This format is illustrated by Table 7-1.

Sometimes a column is added to reflect the significance of the identified failure mode and its immediate consequence upon the system. This may be a qualitative estimate of criticality or a numerical expression of probability of occurrence. This enables the analyst to sort the results and produce a ranking of component failures either by system criticality or by probability of occurrence (see Table 7-2).

DOCUMENTS AND DRAWINGS REQUIRED

The master documents to which other documents should be referred are the piping and instrumentation diagram (P&ID), the electrical one-line diagram (EOL) and, if available, a system description. Also useful are training manuals, vendor manuals where applicable, and system operating procedures. Mechanical failures are not the principal or only failures considered, but they provide an accounting mechanism or road map for the inclusion of other components and failure modes. For example, inclusion of a mechanical component such as a pump would also include its derivative components—the dedicated power circuit breaker, its startup and shutdown logic, associated controls, and so forth.

The next group of documents that should be obtained are instrument logic or ladder diagrams, electrical elementary or wiring diagrams, and instrument loop diagrams. Many significant failure modes of a system are

Table 7-1 Sample FMEA Data Sheet; HF Air Systems Mechanical

Drawing	Component Identifier	Component and Failure Mode	Method of Failure Detection	Effect on System	Other Remarks
HF-A-C	A001AIRM	Line drainers		Wet air	
HF-A-C	A002AAIRM	Plant air System drainers (2 of 2) fail		Wet air	
HF-A-K	A003AIRM	Aftercooler Trap A fails closed		Compressor A fails	
HF-A-G					
HF-A-D					
HF-A-F	A004AIRM	Aftercooler Trap B fails closed		Compressor B fails	
HF-A-G					
HF-A-J	A005AIRM	Aftercooler Trap C fails closed		Compressor C fails	
HF-A-G					
HF-A-H	A006AIRM	Aftercooler Trap D fails closed		Compressor D fails	
HF-A-G					
HF-A-B	A007AIRM	Corrosion failure		Air systems inoperative	Single failure

HF-A-B	A008AIRM	PCV 1039 failure	Air systems inoperative	Single failure
HF-A-B	A009AIRM	PIC 1039 instrument air fails	Air systems inoperative	Single failure
HF-A-F HF-A-K HF-A-J HF-A-H	A010AIRM	Cooling htr P GE400 KPA compressor limit	Air systems inoperative	Singe failure
HF-A-K	A011AIRM	Compressor A lube system fails	Compressor A fails	
HF-A-K	A012AIRM	Compressor A unleader fails open	Compressor A fails	
HF-A-E	A013AIRM	Manual valves closed—1 of 5	Manual valve or cooling water circulation failure	
HF-A-E	A014AIRM	4 of 4 TIA inoperative	4 of 4 TIA failure	

Table 7-2 Failure Modes and Effects Analysis

Item	Failure Modes	Cause of Failure	Possible Effects	Probability of Occurrence	Criticality	Possible Action to Reduce Failure Rate or Effects
Motor case	Rupture	a. Poor workmanship b. Defective materials c. Damage during transportation d. Damage during handling e. Overpressurization	Destruction of missile	0.0006	Critical	Close control of manufacturing process to ensure that workmanship meets prescribed standards. Rigid quality control of basic materials to eliminate defectives. Inspection and pressure testing of completed cases. Provision of suitable packaging to protect motor during transportation.
Propellant grain	a. Cracking b. Voids c. Bond separation	a. Abnormal stresses from cure b. Excessively low temperatures c. Aging effects	Excessive burning rate; overpressurization; motor case rupture during otherwise normal operation	0.0001	Critical	Carefully controlled production. Storage and operation only within prescribed temperature limits. Suitable formulation to resist effects of aging
Liner	a. Separation from	a. Inadequate cleaning of motor case after fabrication b. Use of unsuitable bonding material c. Failure to control bonding process properly	Excessive burning rate; Overpressurization; Case rupture during operation	0.0001	Critical	Strict observance of proper cleaning procedures. Strict inspection after cleaning of motor case to ensure that all contaminants have been removed.

From: W. Hammer, Handbook of System and Product Safety, Prentice-Hall, Englewood Cliffs: NJ, p. 153, 1972 by permission.

often associated with the electrical and control aspect of the system. Depending on how thorough an examination is needed, the last two sets of documents are optional but may provide the only source of important interlocks and system control dependencies, otherwise unaccounted for in higher level documents. It is important to remember that a system description and training materials emphasize how a system is supposed to work, not how it might work in spite of the designer's intent.

The aforementioned documentation will show the complete effect on the subsystem or system of the loss of a component by a certain mode of failure. These effects can be determined only when both the support interactions of "distributive" systems such as instrument air, cooling water, or electrical power, and the logical relationship of the controls, are understood. Where a fault tree (discussed in Chapter 9) makes this explicit, the FMEA must implicitly treat these interactions.

For example, Table 7-1 shows that simple failure of the cooling water flow will cause failure in all four air compressors. Though this is an obvious system interaction, there are more subtle ones, such as failure of the instrument power to the pressure controller PIC 1039, which is not shown on this sheet. The point is that the analyst must have enough basic knowledge of a component's construction and operation to pick up the interactions that are fatal and those that emanate from the system design in which they are embedded.

SELECTION OF TEAM MEMBERS

It is recommended that the lead member of the FMEA study team have the following qualifications for the study to have meaningful results:

- Prior experience with equipment involving broad exposure to the causes and effects of transients and equipment failures
- Knowledge of system engineering involving both controls and mechanical or electrical design
- Ability to organize, train, and administrate a team of analysts using the FMEA technique.

The reason that such broad knowledge and prior experience are needed is that the FMEA method does not generate the detailed functional relationships and dependency information that is produced by more thorough (and more costly) techniques such as fault tree analysis. It follows that what is lacking in the details of the method must be made up in the combined experience of the lead analyst and his team. The rest of the team should therefore include a mix of disciplines and backgrounds that complement each other, since it is rare that one individual would possess all attributes. If the right type of experience is available, only two more members are required:

- A system engineer or operations specialist familiar with the design and operation of the system
- A controls specialist, preferably familiar with both electrical and control design, logic, and equipment used.

It is essential that the team be able to assess not only the immediate effect of a failure on the component, but also on the system operating parameters and the expected responses of the controls. For example, a particular mode of failure of an instrument can simulate a real process transient, initiating a compensating control action, which then produces a more immediate failure of the system to operate.

Pumps are normally protected from closed valves on the suction side by a low-pressure switch. Reciprocating compressors are normally protected from loss of lubrication by a low oil pressure switch and a time delay. In either case, if the switch fails low, the result is the same as if the real process condition existed.

Our previous example of the air compressors (Table 7-1) includes failures of line drainers and traps on the aftercooler, either of which can cause the back up of incompressible fluid into the compressor cylinders, shattering the pistons and destroying connecting rods and journals. It is also not unheard of to have a cooling water pressure that is higher than the air pressure, which in the event of aftercooler tube leaks, could have the effect of introducing more water than the traps and drainers could be expected to handle, and damage to the compressors could result. Therefore, if the compressor is treated as a "package" and the construction, layout, and operation conditions of the subcomponents are ignored, the analysis may not develop important failure modes.

In one plant visited by the author, the air line drainers were in the bottom of a trench, buried under several piping lines and a heavy metal grating. It was not surprising that they had not been maintained or checked since they were first installed. Judging from the fact that the downstream air receiver had corroded through before the plant was even started up, it was doubtful that the drainers had ever worked properly.

STUDY GUIDELINES

System Definition

An overall objective of an FMEA is to organize and document what is known about the effect of component failures on a specified system.

Systems, as they are commonly defined, are an arbitrary or traditional grouping of components. In other words, the system defined on a P&ID is not usually functionally complete in that it depends on outside sources of power, cooling water, or control information. The first job of the analyst is therefore to define the complete functional boundaries of the system to be studied. This is best accomplished by marking up a set of drawings and annotating them to show their functional limits and dependencies.

Level of Detail

The next item to be resolved in conducting an FMEA is to decide what level of detail will produce the results desired. Most FMEAs stop at the component level because the study involves the identification of hazards at the plant or operating unit level and involves several systems. However, for major pieces of equipment, many events are initiated at the subcomponent level. Because this level is not addressed in a typical FMEA, the subcomponents have to be included implicitly in the failure modes of the component as each component is addressed. Subcomponent failures can be included only if the study team has the expertise to include this type of failure.

The overall objective of the analysis determines the level of detail necessary. The problem may be better suited to either a cursory WHAT-IF analysis or a more thorough and detailed fault tree analysis (FTA). Like hazard and operability studies (HAZOP), FMEAs generally fall in between a WHAT-IF and an FTA in level of effort but require fewer resources than a HAZOP, which examines the operating success space as well as its failure space. A HAZOP must examine many trivial, nonthreatening, and otherwise successful cases in addition to those that produce undesired results. "Failure space" methods such as FTA concentrate effort on undesired results only but attempt to go into more depth as to *why* something might occur. Both have their uses.

Data Sheet Design

An FMEA data sheet typically includes

- Identification of the component and parent system
- Failure mode selected; cause of failure
- Effect of the failure on the subsystem or system
- Method of detection; diagnostic aids available
- System and operator response; elucidating comments.

These features are shown in Table 7-1 and are discussed in the following text.

Component identifier. Proper component identification includes three elements:

- A functional title or descriptive (e.g., "methanol feed pump").
- A mark number or identifier that can be tied to the base drawing.
- A parent subsystem or system tying the component to a set of drawings and a system description.

    The latter two identifiers may be coded, but the first item should be in plain English so that the results in the final document are self-evident.

Component and failure mode. Description of the failure mode in a columnar format is necessarily concise, but it must be clear and bound by the nature of the failure. It should also be realistic. Equipment in a similar duty has a typical set of failure modes, which can be ranked in order of their probable occurrence rate. One way to clarify the nature of the failure is to add a column to the data sheet listing the cause of the failure, thereby describing the failure scenario implicitly.

Effect on system. This column requires the most effort and thought of all the headings in the data sheet. It also requires that the failure be examined from a multidisciplinary perspective. This is not only to be certain that mechanical effects on the system are considered, but also to consider the electrical and control effects of the failure impact on the process parameters. Most failure effects on the component itself are straightforward and obvious, but the ability to project the down-stream impact of failures occasionally requires specialized knowledge. Some of this knowledge may reside only in those who are thoroughly familiar with the system's operation, or with those who have broad experience with failure effects of this nature in similar processes. In any case, the team must be able to recognize when this specialized knowledge is needed and to avoid faulty recollection or common folklore as a substitute.

    Frequently information from operators and even plant engineers can be characterized as empirical. It is not based on any intimate knowledge of the equipment design or on whether the equipment is being operated outside the range for which it was originally specified. A specialist or manufacturer's representative with both engineering and service knowledge may be necessary to interpret some component problems.

Method of failure detection. Sometimes the prior detection of anticipated failures poses a difficult problem for the FMEA study team. Frequently instruments that could provide the required information are inoperative or missing. Sometimes the possibility of detection may be masked or obliterated by the response of the controls. Some failures have a long incipient development period or can be detected at an early stage by condition monitoring techniques. In any case, it is imperative that where the analysis takes credit for operator intervention, the operator must have had an unequivocal, clear indication that a failure has occurred or is in process. Failures at an early stage are often subtle and require more effort to identify. Recent attempts to develop diagnostic aids using knowledge bases or expert systems have shown that these require a large investment of time, and sometimes great expense. Because of the number of inherent pitfalls in failure detection, it is essential that the analyst be unbiased in his or her appraisal.

System and operator response. If automatic controls or the system capacity is able to absorb the effects of a specified failure with no loss of system function, then this should be documented. This is an informed judgment, however, and should be backed up by solid evidence or operating experience. Frequently, speed of the operator recovery action in response to a failure is overestimated when the failure either happens too fast for such a response to be effective, or the information available to the operator is inadequate for such a response. Here, the analyst must be realistic about the operator's ability to perform. Resolutions requiring changes in operator participation should also be written into normal or emergency operating procedures, since the analytical results depend on them. This also ensures that such a resolution gets a thorough review before it is implemented.

REVIEW OF DOCUMENTATION

FMEA Evaluation

The first product of an FMEA is a collection of worksheets that show the results of single component failures on a system. Without some indication of the severity of each effect and the likelihood of its occurrence, all such failures that result in a loss of system function are considered equal. However, permanent damage to equipment or long outages may result from some failures. Therefore, it is incumbent upon the analyst to either add a column to the worksheet that identifies the criticality to system function of each failure, or to further characterize the effect of the failure in terms of potential loss of production or revenue in the Comment column under the response heading.

The filled-in worksheet should make it obvious to the reader which single failures disable system function and, from their cause description or a semiquantitative "guesstimate" of probability, which of these single failures is most likely to occur. Design provisions to detect or prevent the initiation of these failures or to mitigate their effects should be evaluated for their effectiveness. If the design change needed is extensive, then a mini-FMEA should be performed on it as well, because it may introduce a new set of hazards and failures.

METHODS FOR DOCUMENTING THE STUDY

The basic document produced in an FMEA is the handwritten worksheet illustrated in Table 7-1. Alternatively, this can be produced as a data base, using commercially available software, and the worksheet design used as a report template. It is also important to record the sources of input data separately, along with the interpretation and conclusions drawn from the evaluation in the previous section.

ADVANTAGES AND DISADVANTAGES

The advantages of the FMEA are as follows:

- Ease of construction at the component level
- Ease of interpretation by the layperson
- Less time taken than for a more detailed study methodology
- Quickly reveals fatal single failures when properly executed.

The disadvantages of the FMEA are as follows:

- Addresses only one component at a time and may not reveal important interactions between components and other components or systems
- Does not develop sufficient detail to provide a uniform basis for quantification of system effects
- For good results, requires high grade experience on the study team.

SUGGESTED READINGS

Henley, E. J., and Kumamoto, H. 1981. *Reliability Engineering and Risk Assessment.* Englewood Cliffs, NJ: Prentice-Hall.

PRA Procedures Guide: A Guide to the Performance of Probabilistic Risk Assessments for Nuclear Power Plants, NUREG/OR 2300, January 1983, Section 3.6.1.

Klaassen, K. B., and van Peppen, J. C. L. 1989. *System Reliability, Concepts and Applications.* New York: Edward Arnold, Routledge, Chapman and Hall, Inc.

8

Hazard and Operability Studies

R. M. Sherrod and W. F. Early

A hazard and operability study (HAZOP) is a simple yet structured methodology for hazard identification. It is a program that allows its user to employ imaginative thinking in the identification of hazards and operational problems.

A HAZOP involves a systematic, methodical examination of design documents that describe the facility. The study is performed by a multidisciplinary team to identify hazards or operability problems that could result in an accident. Deviations from the design value of key parameters are studied, using guide words to control the examination evaluation. This presumes that the design values of flows, temperatures, and other process variables are inherently safe and operable. The study team consists of trained personnel, knowledgeable in the technology and operations and having the necessary technical expertise to answer most questions raised during the review without outside assistance.

The study is used to

- Provide management with knowledge of where potential hazards may exist and to provide a vehicle for later recommendations for plant design or procedural modifications.
- Provide safety-related documentation of every line and piece of equipment in the plant, which is very useful when modifications are carried out.
- Provide a prioritized basis for subsequent risk analysis or assessment work.
- Provide a basis for a risk management program as currently mandated by legislatures in the states of California, Delaware, and New Jersey, and under consideration in several other states. The Occupational Safety and Health Administration (OSHA) is also instituting guide-

lines that would focus on worker issues but would effectively mandate the same type of program.

If a HAZOP is performed and if modifications are subsequently implemented to mitigate risk, then the plant should be operating under some lower risk than before the study. Performing a HAZOP, however, does not guarantee that no risk exists, nor does it guarantee that hazardous events or operating problems will not be encountered.

DISCUSSION

The following discussion is a stepped, methodical review of a HAZOP, its strengths and weaknesses. Also introduced is the basic HAZOP guide word approach as attributed originally to Imperial Chemical Industries, Ltd. (ICI) and further defined by the American Institute of Chemical Engineers—Center for Chemical Process Safety and developed through practical application by end users.

As a starting point, we define the following terms:

Hazard   Anything (chemical reaction, equipment malfunction, or operator error) that can lead to an unwanted event.

Operability   Anything that causes the operator to improvise in his or her actions.

These definitions are an important part of the basic HAZOP premise that the process does not have inherent hazards or operating problems when the unit is operating as designed, and as defined by the basic documents such as the process flow diagrams, equipment specifications, and operating procedures. In other words, if there are no deviations from the norm, there are no problems.

A HAZOP is a simple, formalized methodology to identify and document hazards through imaginative thinking. As mentioned, the methodology was originally introduced by ICI. While a useful and powerful tool when properly used, it can be a waste of time if plant documentation is out of date and there are no plans to update it, or if basic safety requirements are ignored. A HAZOP can be performed at any time during the design or operation of a plant and can provide a truly analytical method for identifying hazards. It is done to find the "weak link" in a plant and to provide a basis for developing procedural or engineering controls to remove or lessen the risks from the identified problem area. During the design stage of a new unit, the HAZOP is generally performed at the time of issue of production quality piping and instrumentation diagrams (P&IDs) when essentially all basic design decisions have been made, but before the design has proceeded too far to incorporate any modifications deemed appropriate. It is also a recommended tool when making plant modifications or revamps, since changes in the basic design introduce opportunities for error that might not be obvious unless reviewed as part of a "system" instead of a localized change. HAZOPs are also very useful for existing facilities, where they can be used for evaluating operations from an unbiased viewpoint, to identify possible process improvements, or as a quality-assurance effort.

Frequently, a HAZOP is performed following an incident or accident that has resulted in a serious substance release or injury. HAZOPs are also often performed at the request of an administrative agency that is exercising its regulatory mandate. Either of these situations generally prompts some sort of hazard identification, typically followed by a hazard analysis of any identified critical hazards. Most often, these situations occur in operating plants with long operating histories rather than in new facilities. In this type of situation, thorough and up-to-date documentation is essential. As stated before, a HAZOP is appropriate for any facility with P&ID-type drawings. The importance of having complete, up-to-date documentation for the HAZOP cannot be overstated, and with this in mind, it is normal and prudent that the HAZOP be performed at the location most likely to have the documentation on hand. During a design phase, this is usually the contractor's office. For modifications or revamps of an existing unit, this documentation is typically at the plant site.

A HAZOP is not an analysis for determining how far one goes in physically removing any risk or mitigating the consequences. It is also not the responsibility of the HAZOP team to come up with the engineering or procedural solutions. These can be accomplished later and would be time-prohibitive if attempted by the team. The team must also keep in mind that hazards may frequently be mitigated by instituting procedural changes rather than changes requiring capital expenditures. Wherever deemed appropriate, the procedural solutions are cost effective and quickly and easily instituted. Frequently, a combination of both procedural and engineering changes is employed in mitigation.

As mentioned earlier, the HAZOP guide word approach focuses on deviations from the normal or design operating parameters. It is intuitively easy to see how this applies to steady-state operations, but it is equally applicable for batch or semibatch operations, since the basic premise that normal operations are inherently safe and that deviations are the source of unrecognized problems remains the same.

As an adjunct to the HAZOP effort, the team often finds the need to review procedures from a critical standpoint instead of just using them as background documents. This procedural review typically addresses such items as startup, shutdown, loading or unloading of acutely hazardous materials (AHMs), and other unusual circumstances that can affect worker or operational safety. A simple review method is to outline the inherent hazards, outline a safe and simple approach, and then review the existing procedure against the simple, idealized procedure. Many things can be learned through this approach, not the least of which is that during procedure preparation, one often overlooks basic problems because of the need for detailed development of key instructions. This simple review quickly assesses the procedure as to its thoroughness and completeness and also educates team members. This educational aspect can be realized most readily by addressing critical procedures at the start of the HAZOP.

The basic HAZOP guide words and their logical extension to operating parameters are as follows:

Guide Word	Operating Deviation	
None	No flow	
	Reverse flow	
	No reaction	
More	Increased flow	
	Increased pressure	
	Increased temperature	
	Increased reaction rate	
Less	Reduced flow	
	Reduced pressure	
	Reduced temperature	
	Reduced reaction rate	
Part of	Change of ratio of materials present	
As well as	Different materials present	
Other	Different plant conditions from normal operation	
	Startup	Erosion
	Shutdown	Severe cold
	Relief	Earthquake
	Instrumentation	Hurricane
	Sampling	Tornado
	Utility failure	Airplane crash
	Corrosion	Flooding
	Maintenance	Sabotage
	Grounding/static	

The use of these guide words and their application within the framework of a HAZOP are discussed in detail later in this chapter. However, several other topics are discussed first, so that the detailed presentation can be seen in the proper perspective. The follow-up to the guide word approach is documentation of the effort that presents the causes and consequences of the operating deviations, the suggested action for the problem (suggested mitigation, which the team deems applicable without spending any additional time on the hazard), criticality and frequency ratings, and hazard ranking. A sample format for recording the HAZOP is shown in Figure 8-1.

A proper mix of meeting participants is very important in the performance of a HAZOP. The mix of participants normally includes the following:

Team leader	HAZOP experience is imperative.
Process/mechanical engineer	Familiarity with P&ID elements is required.
Plant operator	Must have actual knowledge of the operating unit to bring a "real-world" perspective to the study.

Process engineer
Instrumentation engineer
Electrical engineer
Safety or fire protection engineer
Toxicologist
Maintenance personnel
Team recorder

Many of these personnel are often only part-time participants but remain on call for assistance. For the toxicological considerations, electrical, maintenance, and other subjects, participants in these areas may be required only on a consulting basis. Also, it is important to keep the size of the review team to a manageable number, such as between five and seven, so that all members of the team can participate.

In the following sections, we present such aspects for total HAZOP study development as preplanning issues, study logistics, and practical examples. In addition, a recommended checklist of items to be included in the HAZOP report is presented. This checklist can be used in preparation or review of the study to ensure that the technique has been properly used, since regulatory agencies now frequently mandate certification.

STUDY TITLE: _____

UNIT: _____

LINE/EQUIPMENT NUMBER: _____

LINE/EQUIPMENT DESCRIPTION: _____

JOB NUMBER: _____

P&ID NO.: _____ ISSUE: _____ DATE: ___/___/___

BY: _____ DATE: ___/___/___ PAGE: _____ OF _____

GUIDE WORD DEVIATION	POSSIBLE CAUSE	CONSEQUENCES	SUGGESTED ACTION	RANKING		
				C	F	R

C = CRITICALITY   F = FREQUENCY RATING   R = HAZARD RANKING

Figure 8-1   Hazard and operability study action report.

HAZOP PREPLANNING ISSUES

Preplanning issues addressed in a typical refinery unit HAZOP include the following:

- Verification of as-built conditions shown on the P&IDs
- Line segment boundaries set; markup of P&IDs
- List of support documents compiled
  - P&IDs (base study document)
  - Process flow diagrams (PFDs)
  - Process description
  - Operating manuals/procedures
  - Processing materials information
  - Equipment and material specifications
- Tentative schedules of time to be spent per P&ID sheet
- Recording technique (computer program or data sheet) determination
- List of standard abbreviations and acronyms compiled
- Criticality rankings devised
- HAZOP training given to all team members (one day)
- Arrange for system or process briefings for team before work begins.

These preplanning activities are typical of the preparatory steps required to properly implement a HAZOP. The preplanning should be done immediately before the HAZOP is executed. Preplanning efforts frequently take between one day and one week.

HAZOP STUDY LOGISTICS

Logistical development of this refinery unit HAZOP included the following:

- Preplanning issues were addressed the prior week.
- The team included three core team members and four part-time members.
- The study included 16 moderately busy P&IDs.
- The study took three and one-half weeks
- The team met 4 hours per day in morning review sessions and spent 2 hours per day on individual efforts for reviews, follow-ups, and field checks.
- Dedicated space was required for storing the large number of documents.
- The study resulted in 170 data sheets.

- The team recorder used a personal computer to record, sort, and retrieve data. The Stone & Webster proprietary program PCHAZOP[a] was used.
- The plant operator was the key contributing plant member of the team.
- Key operating procedures were reviewed relative to the P&IDs and safe engineering practices.

The keys to success of any HAZOP include completeness and accuracy of the documentation, appropriate technical skills of team members, the ability to use the guide word technique comfortably, and the ability to concentrate on real hazards while avoiding getting sidetracked by peripheral issues. Not all combinations of guide words and parameters represent hazards, and this fact should not be allowed to confuse the issue.

HAZOP AGENDA WITH EXAMPLES

A typical HAZOP agenda is as follows:

- Introduce team members. A short biographical sketch is given for each team member.
- HAZOP methodology. A presentation of the methodology to be used in the study is made by the team facilitator. This also establishes the team member mind-set necessary to conduct an effective HAZOP.
- Properties of acutely hazardous materials (AHMs). An identification and discussion by an industrial hygienist or other knowledgeable person is presented to establish a team recognition of the AHM toxic and hazardous properties.
- Identify hazards. A preliminary identification by team members of hazards present within the process. This serves to focus the HAZOP.
- Purpose or goal of a HAZOP. A general presentation of the purpose or goal of the HAZOP is made that serves to direct the HAZOP.
- Scope the HAZOP. Once the AHM properties are established and discernible HAZOP hazards are identified, the HAZOP scope is set.
- Establish criticality and frequency ratings. Hazard classification matrix parameters are set by the team for subsequent screening of HAZOP identified hazards.
- Define dispersion models. The team identifies AHM release scenarios

---

[a] PCHAZOP is a proprietary, PC-based HAZOP program designed to provide a simple method for recording and documenting the HAZOP process. Stone & Webster's program is designed to run in a data base environment to facilitate data storage and retrieval.

that will be different cases for dispersion modeling to be performed outside the HAZOP team meeting.
- Process description. A detailed process discussion is presented by the process engineer to familiarize the HAZOP team with process scenarios that may lead to a hazardous condition, as well as to ensure that all team members have a clear understanding of the basic process.
- PCHAZOP on-line. Load the PCHAZOP* computer program for recording HAZOP team proceedings.
- Plant walk-through. The team walks through the unit being studied to obtain an overview of the plant facilities.
- HAZOP team meetings. A systematic segment-by-segment team review of the facilities contained within the job scope is executed.
- Interim report. A mini-report providing immediate notification to plant management of any critical hazard that might be identified during the HAZOP team meetings should be developed immediately following completion of the study.
- HAZOP report. A formal report is prepared by the team facilitator, reviewed and commented upon by the other members, and then made a permanent record of the team meetings. Guidelines are presented later in the chapter for report contents.

We will now expand upon the preceding agenda, using an example problem as a basis for discussion. The example is shown in Figure 8-2 and represents a simplified process scheme for producing ethylene glycol from ethylene oxide and water. In this case, ethylene oxide is the primary focus because it is an acutely hazardous material.

Following the initial steps of introduction of team members and presentation of the HAZOP methodology, a hygienist presents a synopsis of the toxic and hazardous properties of ethylene oxide. A simple presentation includes the following (typically, this presentation is quantitative, but it will be presented qualitatively here):

- Toxicity data
- Flammability limits
- Explosive limits
- Chemical stability data
- Chemical reactivity data.

Next, the preliminary hazard identification is performed by team members identifying hazards present within the process. For the ethylene oxide example, the hazards are presented as follows (again, qualitatively):

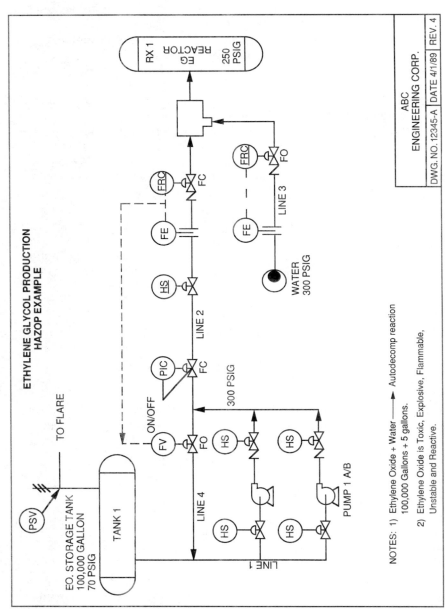

Figure 8-2 Ethylene glycol production HAZOP example.

- Fire hazard
- Explosion hazard (both confined and unconfined)
- Toxic exposure hazard
- Missile fragments (secondary effect of explosion hazard)
- Chemically reactive.

Once the acutely hazardous material properties and other hazards have been identified, the purpose (or goal) and the scope of the HAZOP have been set. Then the question of assigning criticality and frequency ratings must be addressed. Classification matrix parameters are set by the team for subsequent screening of HAZOP-identified hazards. Criticality ratings may be assigned by the HAZOP team as deemed appropriate for the HAZOP being addressed. A common, basic example of assigned ratings is as follows:

CRITICALITY RATING

I	Catastrophic	Results in either
		• An on-site or off-site death
		• Damage and production loss greater than $1,000,000
II	Severe	Results in either
		• Multiple injuries
		• Damage and production loss between $100,000 and $1,000,000
III	Moderate	Results in either
		• A single injury
		• Damage and production loss between $10,000 and $100,000
IV	Slight	Results in either
		• No injuries
		• Damage and production loss less than $10,000

Similarly, frequency ratings are assigned by the team on a generalized basis. An example follows:

FREQUENCY

A. Occurs more than once per year
B. Occurs between 1 and 10 years
C. Occurs between 10 and 100 years

   D. Occurs between 100 and 10,000 years
   E. Occurs less often than once per 10,000 years

(Note that frequencies are separated by orders of magnitude to assist in interpretation if a fault tree or other truly quantitative analysis is performed later.)

A risk matrix is then prepared (as shown in Figure 8-3) to provide for a further ranking of identified hazards on the basis of acceptability or, at a minimum, to provide for a more easily sorted and prioritized ranking. Figure 8-3 is a sample risk matrix that is a fair representation of industry practice.

Risk rankings have been assigned within the risk matrix to provide a quick and simple priority sorting method. The rankings are then given definitions that include, if the facility operator desires, definitions and recommended actions similar to those in Table 8-1.

This approach is used primarily to focus on hazards that represent either worker or societal risks and are therefore the primary concern. However, many of the items given criticality rating IV, and especially those with high frequency ratings, should be reviewed closely to determine the effect of mitigatory steps, since these are the findings that most often produce a

		FREQUENCY				
		A	B	C	D	E
CRITICALITY	I	1	1	1	2	4
	II	1	2	3	3	4
	III	2	3	4	4	4
	IV	4	4	4	4	4

Figure 8-3   Risk matrix.

Table 8-1   Definitions and Recommended Actions for Rankings

Ranking	Description	Required Mitigation
1	Unacceptable	Should be mitigated with engineering or administrative controls to a risk ranking of 3 or less within a specified time period such as 6 months.
2	Undesirable	Should be mitigated with engineering or administrative controls to a risk ranking of 3 or less within a specified time period such as 12 months.
3	Acceptable with controls	Should be verified that procedures or controls are in place.
4	Acceptable as is	No mitigation action required.

Note: For any hazard assigned a risk ranking of 1 derived from a criticality/frequency rating of I-A (most critical event), the HAZOP team should perform a thorough multiple-cause investigation to identify all possible causes for input to the mitigation step.

tangible payoff from the HAZOP and can justify the HAZOP effort financially.

Next, dispersion models are defined and studies are done to calculate the spill sizes that correspond to criticality rating codes. This is required to ascertain what size spill could result in an off-site death, what size spill could result in an on-site death, what size spill could result in an on-site injury, and so forth. For instance, could a drum-size spill result in an off-site death or an on-site death? This determination is necessary to assign criticality ratings during the HAZOP.

Example

At this point, the team is nearing the start of the study, has assembled at the plant site or other appropriate study venue, and is ready for the detailed process discussion by a process engineer. A simplistic presentation of the example problem would proceed as follows:

Ethylene oxide (EO) and water are combined in a simple, ratio-controlled mixture to produce ethylene glycol (EG) (Figure 8-2). For this study, the scope begins with the ethylene oxide storage tank and ends with the ethylene oxide-water reactor. The reaction itself is exothermic, but the water-to-EO ratio is excessive so that the water can provide a heat sink for the heat of reaction. Reversing the ratio will result in violent reactions and EO autodecomposition with as little as 5 gallons of water in a 100,000-gallon EO storage tank.

For the purposes of this study, the EG process begins at the EO storage tank.

This is a pressurized vessel designed for 125 psig and provided with a relief valve (RV) having that set pressure. The RV is designed for simple overpressure cases only, since an EO autodecomposition reaction proceeds at such a rapid rate that no RV can relieve sufficiently to protect the vessel. The RV relieves to a relief header, which will not be part of this study.

From the storage vessel, EO is pumped using a dual pump set through piping and flow-control valves (Figure 8-2) to the reaction vessel. Before entering the reactor vessel, it combines with the larger water stream and proceeds through a static, in-line mixer. The EO piping contains a back-pressure control valve for backflow prevention and a flow-control valve to control the EO flow. In the event of too little EO flow, there is an on-off flow control valve in a kickback line to ensure that the pump always operates at or above its minimum flow. The kickback line was not designed for continuous kickback, because the constant energy input could heat the EO and potentially contribute to autodecomposition.

The EO pumps are set up only for single pump operation. No autostart capabilities are provided, since a blocked-in pump could result in autodecomposition of EO. Hand-actuated, motor-operated valves are provided to isolate each pump, and these are actuated by the outside operator. Additional hand-actuated valves and check valves are provided as deemed appropriate throughout the process.

In a real setting, several additional valves, gauges, ratio control tie-ins, and other controls would be provided. Also, the pumps would be provided with a closed vent and drain for safe drainage and disposal of EO. For the purpose of this example study, these items are unnecessary.

The next item on the HAZOP agenda is to ready the method for recording the HAZOP team proceedings for use. The HAZOP record may be kept either manually or on a computer disk. At this point, if the HAZOP record is to be kept on a computer disk, the team recorder sets up the computer to record the HAZOP team proceedings. The computer model allows retrieval and sorting of data to aid in report preparation. The program can also help facilitate the HAZOP sessions by sorting outstanding action items to prepare an action item punch list. Other such programs are available.

Plant Walk-Through

Next is the plant walk-through. The team takes a short walk through the unit being studied for an overview audit of the plant facilities. This is done at the beginning or during the early stages of the HAZOP to obtain a general feeling for the condition of unit housekeeping equipment (e.g., the emergency response equipment, sensors, alarms) maintenance access, operator access, and valve manifold complexity. This aids the team in assessing hazard criticality and probability when conducting the HAZOP.

At this point, the team is ready to perform the systematic segment-by-segment team review of the facilities contained within the job scope. Examples of the type of information reviewed and the resulting entries—usually in handwritten report form—are shown in Figure 8-4. Similar entries are presented as recorded in PCHAZOP* and printed on a desktop printer (Figures 8-5 and 8-6).

The line chosen for review is line 1. A review of the pumps is also presented. Each of these is clearly identified in Figure 8-7 to clarify the segments being studied. It should be apparent that the segments are not always single lines, but are lines or items clearly identified as having a commonality of purpose or function. For example, an in-line static mixer can be part of the line segment and does not require consideration as a separate segment.

Whether the HAZOP record is maintained on handwritten forms or on a computer disk, the team recorder must resort to shorthand notation and abbreviations to conserve time and space. The HAZOP report must be concise, but not to the extent of being incomprehensible. It should be more expansive for the more critical identified hazards. This is relatively easy to accomplish if the report is prepared immediately following the HAZOP team study. Because of the time commitment of HAZOP team members, lengthy HAZOPs are often divided into discrete, week-long sessions. If this is done, the team leader should prepare the report in segments corresponding to the team meeting segments. This helps the team leader in report preparation because details are recorded and not forgotten between meetings. Additionally, topics that are vague or for which differing opinions exist among members can be identified for resolution by the team at the beginning of the next session.

Immediately following completion of the actual HAZOP meetings, a mini-report is prepared. The HAZOP facilitator writes the interim report, which serves as a formal means to disclose any critical identified hazards to plant management and to allow management to take immediate action. This allows management to mitigate any known critical hazards before the final HAZOP report is issued.

INTERIM REPORT

A typical interim report should cover the following items:

- HAZOP identification
- List of team members

STUDY TITLE: MOJAVE ANTIFREEZE CORPORATION

UNIT: ETHYLENE GLYCOL PRODUCTION

LINE/EQUIPMENT NUMBER: LINE SEGMENT 1

LINE/EQUIPMENT DESCRIPTION: PUMP SUCTION LINE

STONE & WEBSTER JOB NUMBER:

P&I NO.: 18426-A   ISSUE: 4   DATE: 4/1/89

BY: MCL   DATE: 6/2/89   PAGE: 1 OF 3

GUIDE WORD DEVIATION	POSSIBLE CAUSE	CONSEQUENCES	SUGGESTED ACTION	RANKING		
				C	F	R
NO FLOW	PUMP TRIP OR MOV SHUT OR LOW LEVEL IN TANK TK-1	LOSS OF PRODUCTION	1. CONSIDER ADDITION OF LOW FLOW ALARM TO FRC TO MINIMIZE PRODUCT LOSS. 2. CONSIDER LOW MOTOR AMPS ALARM.	IV	A	4
REVERSE FLOW	PUMP TRIPS & PIC FAILS/ LEAKS & HS NOT CLOSED/ LEAKS. (NOTE: FV GOES WIDE OPEN)	WATER IN TANK, AUTODECOMPOSI- TION REACTION, CATASTROPHIC TANK FAILURE	SUGGEST COMPLETE BACKFLOW INSTRUMENTATION REVIEW TO DETERMINE ACCEPTABLE METHOD FOR BACKFLOW PREVENTION.	I	B	1
REDUCED FLOW	PARTIAL VALVE CLOSURE	REDUCED PRODUCTION. NOT A HAZARD	SEE NO FLOW	IV	B	4
AS WELL AS	WATER BACKFLOW. SAME AS REVERSE FLOW.					
	NO OTHER IDENTIFIED HAZARDS.					

C = CRITICALITY   F = FREQUENCY RATING   R = HAZARD RANKING

Figure 8-4   Hazard and operability study action report.

STUDY TITLE: __MOJAVE ANTIFREEZE CORPORATION__   STONE & WEBSTER JOB NUMBER: _____

UNIT: __ETHYLENE GLYCOL PRODUCTION__

P&I NO.: __18426-A__   ISSUE: __4__   DATE: __4/1/89__

LINE/EQUIPMENT NUMBER: __PUMP 1 A/B__

BY: __MCL__   DATE: __6/2/89__   PAGE: __2__ OF __3__

LINE/EQUIPMENT DESCRIPTION: __ETHYLENE OXIDE FEED PUMPS__

GUIDE WORD DEVIATION	POSSIBLE CAUSE	CONSEQUENCES	SUGGESTED ACTION	RANKING		
				C	F	R
NO FLOW	DISCHARGE MOV CLOSED	PUMP DEADHEAD, HEAT UP TO AUTODECOMPOSITION TEMPERATURE, CATASTROPHIC PUMP FAILURE.	1. CONSIDER LOW FLOW SHUTDOWN SWITCH. 2. CONSIDER ADDITION OF A COOLER ON THE KICKBACK LINE 3. CONSIDER HIGH TEMPERATURE SHUTDOWN SWITCH. 4. CONSIDER LOW AMPS SHUTDOWN SWITCH.	I	C	1
REVERSE FLOW	PUMP TRIPS & PIC FAILS/ LEAKS & HS NOT CLOSED/ LEAKS. (NOTE: FV GOES WIDE OPEN)	WATER IN TANK, AUTODECOMPOSITION REACTION, CATASTROPHIC TANK FAILURE.	SUGGEST COMPLETE BACKFLOW INSTRUMENTATION REVIEW TO DETERMINE ACCEPTABLE METHOD FOR BACKFLOW PREVENTION.	I	C	1
INCREASED TEMPERATURE	EXTERNAL HEAT SOURCE, SUNSHINE ON OFFLINE PUMP?	HEAT UP TO AUTODECOMPOSITION TEMPERATURE, CATASTROPHIC PUMP FAILURE.	1. VERIFY ISOLATION PROCEDURE OF THE STANDBY PUMP. 2. VERIFY INITIATION TEMPERATURE OF THE AUTODECOMPOSITION REACTION.	I	C	1

C = CRITICALITY   F = FREQUENCY RATING   R = HAZARD RANKING

Figure 8-4 (*Continued*)

117

STUDY TITLE: __MOJAVE ANTIFREEZE CORPORATION__

STONE & WEBSTER JOB NUMBER: _____

UNIT: __ETHYLENE GLYCOL PRODUCTION__

P&I NO.: __18426-A__   ISSUE: __4__   DATE: __4/1/89__

LINE/EQUIPMENT NUMBER: __PUMP 1 A/B__

BY: __MCL__   DATE: __6/2/89__   PAGE: __3__ OF __3__

LINE/EQUIPMENT DESCRIPTION: __ETHYLENE OXIDE FEED PUMPS__

GUIDE WORD DEVIATION	POSSIBLE CAUSE	CONSEQUENCES	SUGGESTED ACTION	RANKING C	F	R
REDUCED FLOW	SUCTION MOV PINCHED	CAVITATION, POSSIBLE AUTODE-COMPOSITION REACTION, AND CATASTROPHIC PUMP FAILURE	VERIFY AUTODECOMPOSITION ENVELOPE			
REDUCED PRESSURE	LOW LEVEL IN TANK 1 OR SUCTION MOV PINCHED	SEE REDUCED FLOW	1. VERIFY AUTODECOMPOSITION ENVELOPE. 2. REVIEW THE TANK LEVEL INSTRUMENTATION FOR ADEQUACY.			
MAINTENANCE	IMPROPER MAINTENANCE PREPARATION & IMPROP-ER PROTECTIVE GEAR	RESIDUAL EO IN PUMP CASE, TOXIC GAS RELEASE, CATASTRO-PHIC PERSONNEL EXPOSURE	1. VERIFY MAINTENANCE PRE-PARATION PROCEDURE. 2. VERIFY PROTECTIVE GEAR IS WORN DURING MAINTEN-ANCE.	I	D	2
	NO OTHER IDENTIFIED HAZARDS					

C = CRITICALITY   F = FREQUENCY RATING   R = HAZARD RANKING

Figure 8-4 (*Continued*)

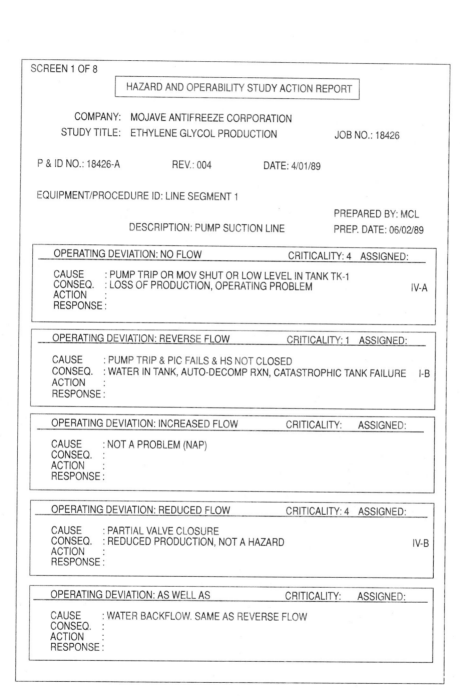

HAZARD AND OPERABILITY STUDY ACTION REPORT

COMPANY: MOJAVE ANTIFREEZE CORPORATION
STUDY TITLE: ETHYLENE GLYCOL PRODUCTION   JOB NO.: 18426

P & ID NO.: 18426-A   REV.: 004   DATE: 4/01/89

EQUIPMENT/PROCEDURE ID: LINE SEGMENT 1

PREPARED BY: MCL
DESCRIPTION: PUMP SUCTION LINE   PREP. DATE: 06/02/89

OPERATING DEVIATION: NO FLOW   CRITICALITY: 4   ASSIGNED:

CAUSE   : PUMP TRIP OR MOV SHUT OR LOW LEVEL IN TANK TK-1
CONSEQ.   : LOSS OF PRODUCTION, OPERATING PROBLEM   IV-A
ACTION   :
RESPONSE :

OPERATING DEVIATION: REVERSE FLOW   CRITICALITY: 1   ASSIGNED:

CAUSE   : PUMP TRIP & PIC FAILS & HS NOT CLOSED
CONSEQ.   : WATER IN TANK, AUTO-DECOMP RXN, CATASTROPHIC TANK FAILURE   I-B
ACTION   :
RESPONSE :

OPERATING DEVIATION: INCREASED FLOW   CRITICALITY: ASSIGNED:

CAUSE   : NOT A PROBLEM (NAP)
CONSEQ.   :
ACTION   :
RESPONSE :

OPERATING DEVIATION: REDUCED FLOW   CRITICALITY: 4   ASSIGNED:

CAUSE   : PARTIAL VALVE CLOSURE
CONSEQ.   : REDUCED PRODUCTION, NOT A HAZARD   IV-B
ACTION   :
RESPONSE :

OPERATING DEVIATION: AS WELL AS   CRITICALITY: ASSIGNED:

CAUSE   : WATER BACKFLOW. SAME AS REVERSE FLOW
CONSEQ.   :
ACTION   :
RESPONSE :

Figure 8-5   Hazard and operability study action report.

| HAZARD AND OPERABILITY STUDY ACTION REPORT |

COMPANY: MOJAVE ANTIFREEZE CORPORATION

STUDY TITLE: ETHYLENE GLYCOL PRODUCTION   JOB NO.: 18426

P & ID NO.: 18426-A   REV.: 004   DATE: 4/01/89

EQUIPMENT/PROCEDURE ID: PUMP 1 A/B

PREPARED BY: MCL

DESCRIPTION: ETHYLENE OXIDE FEED PUMPS   PREP. DATE: 06/02/89

OPERATING DEVIATION: NO FLOW   CRITICALITY: 1   ASSIGNED:

CAUSE   : DISCHARGE MOV CLOSED
CONSEQ.   : PUMP DEADHEAD, HEAT TO AUTO-DECOMP, CATASTROPHIC PUMP FAILURE   I-C
ACTION   :
RESPONSE :

OPERATING DEVIATION: REVERSE FLOW   CRITICALITY: 1   ASSIGNED:

CAUSE   : PUMP TRIP & CKVLV FAILS & PIC FAILS & HS NOT CLOSED
CONSEQ.   : WATER IN TANK, AUTO-DECOMP RXN, CATASTROPHIC TANK FAILURE   I-C
ACTION   :
RESPONSE :

OPERATING DEVIATION: INCREASED TEMPERATURE   CRITICALITY: 1   ASSIGNED:

CAUSE   : EXTERNAL HEAT SOURCE, SUNSHINE ON OFFLINE PUMP?
CONSEQ.   : HEAT TO AUTO-DECOMP TEMP, CATASTROPHIC PUMP FAILURE   I-C
ACTION   : VERIFY 1) STANDBY ISOLATION PROCEDURE, 2) DECOMP INITIATION TEMP
RESPONSE :

OPERATING DEVIATION: REDUCED FLOW   CRITICALITY: ASSIGNED:

CAUSE   : SUCTION MOV PINCHED
CONSEQ.   : CAVITATION, POSSIBLE AUTO-DECOMP RXN, CATASTROPHIC PUMP FAILURE
ACTION   : VERIFY AUTO-DECOMP PHASE ENVELOPE
RESPONSE :

OPERATING DEVIATION: REDUCED PRESSURE   CRITICALITY: ASSIGNED: DIZ

CAUSE   : LOW LEVEL TANK 1 OR SUCTION MOV PINCHED
CONSEQ.   : SEE REDUCED FLOW
ACTION   :
RESPONSE :

OPERATING DEVIATION: MAINTENANCE   CRITICALITY: 2   ASSIGNED:

CAUSE   : IMPROPER MAINTENANCE PREPARATION & IMPROPER PROTECTIVE GEAR
CONSEQ.   : RESIDUAL EO IN PUMP CASE, CATASTROPHIC TOXIC EXPOSURE   I-D
ACTION   :
RESPONSE :

Figure 8-6   Hazard and operability study action report.

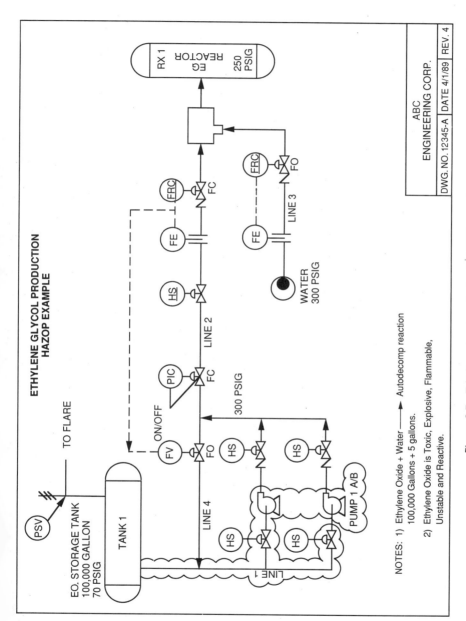

Figure 8-7 Ethylene glycol production HAZOP example.

121

- List of reference documentation
- Form used during review session
- List of preliminary identified hazards
- Basis of risk matrix
- List of base study documents reviewed
- Forms depicting identified hazards ranked 1
- Forms depicting identified hazards ranked 2.

The identified hazards ranked 1 or 2 should be reported on a standard form such as the one in Figure 8-8.

If the HAZOP team identifies a hazard that poses imminent danger to life or property, the team presents its findings to plant management verbally to expedite any required mitigation. Other hazards are identified in the interim report and are further defined in the final HAZOP report.

FINAL HAZOP REPORT

The final effort of the HAZOP is the preparation of a formal report by the team facilitator for a permanent record of the meetings. The report must include the necessary documentation to demonstrate that the basic steps were followed. Laws in some states allow an administrative agency to audit the HAZOP up to five years from the date of study completion, so detailed documentation is necessary. The same laws can prescribe periodic reviews as well, resulting in the need to perpetually update documentation.

Although a HAZOP may have included all the requisites of executing a hazard identification study, no one can judge whether the HAZOP was performed properly unless the information is sufficiently documented in the HAZOP report. Very few questions can arise as to the HAZOP's adequacy if the HAZOP report contains the proper documentation.

The following elements should be included and documented in the HAZOP report for satisfaction of record maintenance requirements and for proper application and implementation of HAZOP hazard identification methodology:

- Purpose and scope defined
- Key members present
- Documentation available
- Proper form used
- Thorough questioning
- Open participation
- Preliminary hazard identification
- Criticality evaluation

MOJAVE ANTIFREEZE CORPORATION	HAZOP
ETHYLENE GLYCOL PLANT	SUMMARY OF CRITICAL
APRIL 1, 1989	IDENTIFIED HAZARDS

CRITICAL RANK: 1, Critical must be improved
                       I-B, I = Potential Fatalities
                          B = Occurs between 1 and 10 years

DEVIATION: Reverse flow

EQUIPMENT NUMBER: Line segment 1

EQUIPMENT DESCRIPTION: Pump (P1 A/B) suction line

CAUSE: Pump trips and PIC valve fails to shut (provide tight
                       shutoff) and board mounted HS fails to shut (provide
                       tight shutoff)

CONSEQUENCE: More than 5 gallons of water backflows into storage
                       tank, initiating auto decomposition reaction, causing
                       catastrophic tank failure.

SUGGESTED MITIGATION: 1. Add primary sensor to close board mounted HS
                          valve on low pressure.
                       2. Verify all backflow prevention devices are tight
                          shutoff type.
                       3. Add primary sensor to shut FRC valve on low flow.
                       4. Consider complete review of backflow instrumentation
                          to determine best method for flowback prevention.
                       5. Perform hazard analysis to determine if final mitigation
                          steps reduce frequency to an acceptable level.

RESOLUTION: All suggested mitigation actions were done including
(To be added in follow up) interconnecting the primary sensors to provide
                       redundant backflow protection.

Figure 8-8   Summary of critical identified hazards.

- Documentation used
- Procedures addressed
- Acutely hazardous material, material safety data sheet addressed
- Spill size quantified and evaluated
- Instrumentation addressed
- Documentation up-to-date
- Thorough cause investigation
- Additional items.

The following is a detailed discussion of the previous checklist items:

- Define purpose and scope
  The HAZOP report should include a statement as to the purpose and the scope of the HAZOP.
- Key members present
  The HAZOP report should include a list of participants in the team meetings, their qualifications, and the attendance roster.
- Documentation available
  The HAZOP report should list the documentation that was the basis for the team review. The list should also include reference documentation available outside the team meeting required to support the study.
- Proper form used
  The HAZOP team data should be recorded on standardized forms containing the key question and answer elements of the HAZOP; one form for each line segment investigated, one form for each equipment item investigated.
- Thorough questioning
  All segments of the HAZOP forms should be filled out to indicate that the HAZOP team executed a thorough study of the system under investigation.
- Open participation
  Although open participation is desirable, this can be evaluated only by actual observation of the HAZOP team in action.
- Preliminary hazard identification
  The HAZOP report should include the preliminary list of hazards identified at the onset of the HAZOP.
- Criticality evaluation
  The HAZOP report should include the basis for the screening matrix that was used during the HAZOP.
- Documentation used
  Although this can be evaluated only by actual observation of the HAZOP team in action, the documentation assimilated during pre-planning should be used in the study.
- Procedures addressed
  The HAZOP report should include a list of the procedures reviewed, including their revision number and date.
- AHM Material Safety Data Sheet (MSDS) addressed
  The HAZOP report should include a list of the acutely hazardous materials, their hazardous properties, and a statement that these were included within the scope of the HAZOP.

- Spill size quantified or evaluated
  The HAZOP report should include a list of any spill size quantification and consequence evaluation that was prepared for use as a basis for criticality ranking during the study.
- Instrumentation addressed
  The HAZOP report should include ample references to the examination of the effects of instrumentation to establish criticality rankings and probability rankings of identified hazards.
- Documentation up-to-date
  The HAZOP report should include a list of revision numbers and issue dates of the actual documents used as the HAZOP study basis.
- Thorough cause investigation
  For critically identified hazards the HAZOP, and hence the report, should include a multiple-cause investigation for use in mitigation efforts.
- Additional items
  The HAZOP report should have a list of abbreviations and acronyms added for clarification.

SUGGESTED READINGS

Chemical Industries Association. 1977. *A Guide to Hazard and Operability Studies*. London: Alembic House.

Elliot, D. M., and Owen, J. M. 1968. Critical examination in process design. *Chemical Engineer,* London, 223, CE377.

Gibson, S. B. 1980. Hazard analysis and risk criteria. *Chemical Engineering Progress* 76(11).

Helmers, E. N., and Schaller, L. C. 1982. Calculated process risks and hazards management. *Plant/Operations Progress* 1(3).

Himmelblau, D. M. 1978. *Fault Detection and Diagnosis in Chemical and Petrochemical Processes*. Amsterdam: Elsevier.

Imperial Chemical Industries Limited. 1974. *Hazard and Operability Studies, Process Safety Report 2*. London: ICI, Ltd.

Kletz, T. A. 1985. Eliminating potential process hazards. *Chemical Engineering.* 92(7):pp. 48–68.

Kletz, T. A. 1983. HAZOP and HAZAN—Notes on the identification and assessment of hazards. London: Institution of Chemical Engineers.

Knowlton, R. E. 1982. *An introduction to guide word hazard and operability studies*. Paper read at CSChE Conference. Montreal, Canada.

Knowlton, R. E. 1982. *An introduction to creative checklist hazard and operability studies*. Paper read at CSChE Conference. Montreal, Canada.

Lawley, H. G. 1974. Operability studies and hazard analysis. *Chemical Engineering Progress* 70(4).

Lercari, F. A. 1988. *Risk Management and Prevention Program Guidance Draft.* State of California Governor's Office of Emergency Services.

Rushford, R. 1977. Hazard and operability studies in the chemical industries. *Transactions N.E. Coast Institute Engineers and Shipbuilders* 93(5).

State of California. 1987. Risk Management and Prevention Program. Chapter 8.95 of Division 20 of the Health and Safety Code.

Solomon, C. H. 1983. The Exxon chemicals method of identifying potential process hazards. Institution of Chemical Engineers. *Loss Prevention Bulletin* No. 52.

Stone & Webster Engineering Corporation. 1988. Hazard & Operability (HAZOP) Study Training Seminar, Houston, TX.

The Center for Chemical Process Safety. 1985. *Guidelines for hazard evaluation procedures.* New York: American Institute of Chemical Engineers.

MIL-STD 882C, System Safety Program Requirement, U.S. Air Force Safety Center, Andrews Air Force Base, Washington, D.C.

9

Fault Tree and Event Tree Analysis

Harris R. Greenberg and Barbara B. Salter

CHARACTERISTICS OF FAULT TREE ANALYSIS AND EVENT TREE ANALYSIS

The preceding chapters of this book discussed various hazard identification techniques. Those techniques are intended as tools for identifying *what* undesired event can occur. Fault Tree Analysis (FTA) and Event Tree Analysis (ETA) are logic modeling tools that are used after one determines the *what* to help you find out *how* an undesired event can happen. FTA and ETA are also tools that can be used, along with component failure rate data and human reliability data, to determine *how often* such an event can occur.

Fault Tree Analysis

FTA is an analysis tool that uses deductive reasoning and graphical diagrams showing the logic of the deductive reasoning process to determine how a particular undesired event can occur. It is a very structured and systematic method that can be used on a single system. It can also be used to model the interaction of multiple systems, accounting for system interactions. It is one of the few tools that can adequately treat the issue of common cause failures, and it is a technique that can produce quantitative as well as qualitative results.

This chapter provides a basic introduction to FTA and examples of how it is applied. A much more extensive treatment of FTA, including a discussion of probability theory and the mathematics of logic (Boolean algebra), is provided in the *Fault Tree Handbook*[1].

Fault Tree Analysis is a technique developed in the 1960s by Bell Laboratories during the Polaris Missile project. First applied in the aerospace industry, FTA was later applied in the nuclear power industry and in the chemical process industries.

The technique can be very rigorous and time-consuming to apply, depending on the level of detail and the extent of the system being modeled. It is probably one of the most intimidating techniques discussed so far in this book, but with the aid of any of a number of computer programs on the market today, fault tree construction and analysis is easier to do and easier to understand than ever before.

The converse of Fault Tree Analysis is success tree analysis, which is used to determine ways to achieve a particular desired event. The approach is very similar, but there are some significant differences in success logic versus failure logic, as we shall soon see.

Event Tree Analysis

Event Tree Analysis (ETA) is a method used to portray an accident as a sequence of events, with various systems or operator actions that can succeed or fail, with consequences of varying severity. It is a technique that is often used in conjunction with FTA to perform quantitative risk assessments.

Event trees are comparable in form to decision trees, originally developed to help business executives choose between alternative plans of action in spite of the uncertainty of future events[2]. Event trees are used to represent a spectrum of possible outcomes of a sequence of events, each of which has an effect on the initiating event or action.

ETA can be used to portray sequences of possible precursors to an accident (such as level detectors or alarm system failures prior to reaching critical reactant concentration or temperature levels). In this case, a spectrum of possible accidents or near-misses can be defined. Event trees can also be used to portray post-accident events (such as release detection, scrubber system actuation, or operator action to close a valve). In the post-accident application, event trees help define a spectrum of possible accidents of varying severity.

Event trees are also used in human reliability analysis (discussed in Chapter 14) to analyze the expected success or failure of someone to follow a given sequence of tasks.

"AND/OR" LOGIC

Logic modeling defines the logic relationships between simple events that combine to cause other events. An example of a familiar logic relationship is the fire triangle shown in Figure 9-1.

Three events must combine for a fire to occur: There must be sufficient

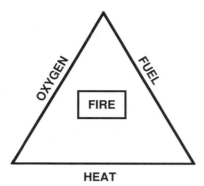

Figure 9-1   "And/Or" modeling: The fire triangle. This model defines the logic relationship between events that combine to cause a different event to occur.

fuel, sufficient air, and sufficient heat. To prevent or to stop a fire, any one of the three ingredients can be removed. That is, take away the fuel, remove the air, or remove the heat source, and the fire is extinguished. This simple logic relationship between the basic elements necessary for combustion helps to quickly identify options for fighting a fire.

Another example, familiar to many, is the following:

Car won't start	Battery dead *or* starter motor shorted *or* ignition switch problem *or* gas tank empty *or . . . or* driver lost key.
or	Any of the contributing events can cause the unde-sired event.
and	All of the contributing events must occur to cause the undesired event.

The problem is that of diagnosing why the car won't start. It could be that the battery is dead, or the starter motor is shorted, or there is a problem with the ignition switch, or even that the gas tank is empty. Any one of these events could lead to the undesired event of the car not starting. In the example of the fire triangle, all of the basic events needed to be present to ensure the top event (i.e., that the fire would start). But in the example of the car, any of the basic events could lead to the top event (i.e., the car not starting).

Simple "and/or" modeling can be developed to describe equipment or situations that involve fairly complex logic. Figure 9-2 shows a device of some sort that processes input signals from three separate sources (signal

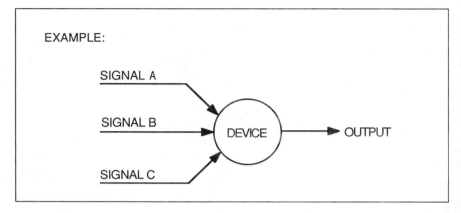

Figure 9-2   "And/Or" modeling: Logic device example. Combinations may be developed for more complex logic.

A, signal B, and signal C). The device has internal logic circuits or decision mechanisms and produces an output signal only if two of the three inputs are present. This complex situation can be described in simple terms with just "and" and "or" logic. That is, output occurs if signal A *and* signal B occur; *or* if signal B *and* signal C occur; *or* if signal A *and* signal C occur.

The description of the logical operation of the device is an arrangement of terms describing the input signals and connecting the possible combinations with the words *and* and *or*. The words *and* and *or* are called logic operators, and they have special symbols that represent them. (These will be described later.)

SUCCESS VERSUS FAILURE LOGIC

To determine how a system can fail, we must first define what the system requires to succeed. To determine the arrangement of logic operators that describe a particular situation, we must consider how success is defined for the system. Figure 9-3 shows a portion of a piping system with four pumps connected in parallel. We must ask ourselves how many pumps are required to be running to meet the system flow requirements. In this case, at least two pumps are required to provide sufficient flow. Two pumps can fail, and the system will still operate successfully, but if three or more pumps fail, the system will fail.

The logic arrangement for system failure based on our definition of system success as "at least two pumps operating," is shown below. The list of possibilities describing system failure is the list of the possible combinations of three out of four pumps failed. The logic arrangement is

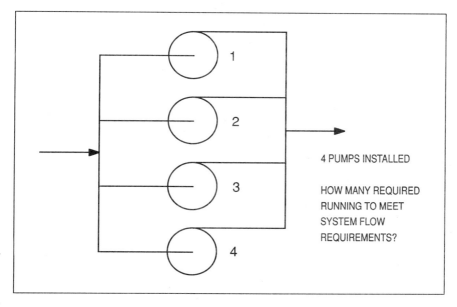

Figure 9-3   Success criteria: Four-pump example.

1. When any combination of three pumps fails out of four installed, the
   system fails, *or*

2. System fails if   Pump 1 *and* pump 2 *and* pump 3 fail
                              or
                        Pump 2 *and* pump 3 *and* pump 4 fail
                              or
                        Pump 1 *and* pump 3 *and* pump 4 fail
                              or
                        Pump 1 *and* pump 2 *and* pump 4 fail.

Success logic and failure logic are the inverse of one another, and the
logic description of one situation is easily translated to the logic description of the other. Also, although not immediately obvious, the description
of one success or failure in terms of logic operators can be converted
directly into the other three steps:

1. Change all of the ANDs to ORs.
2. Change all of the original ORs to ANDs.
3. Change basic event failures to basic event successes (in this case,
   "Pump 2 fails" translates as "Pump 2 runs").

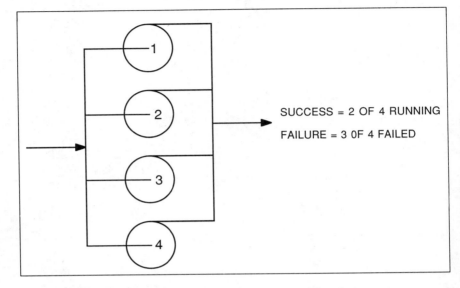

Figure 9-4   Success criteria: Success vs. failure logic.

These steps have been taken to produce Figure 9-4. The translation in this case looks somewhat peculiar, and although the two descriptions do not look alike, they are logically equivalent.

System fails if   Pump 1 *and* pump 2 *and* pump 3 fail
                             or
                       Pump 2 *and* pump 3 *and* pump 4 fail
                             or
                       Pump 1 *and* pump 3 *and* pump 4 fail
                             or
                       Pump 1 *and* pump 2 *and* pump 4 fail

System succeeds if   Pump 1 *or* pump 2 *or* pump 3 runs
                             and
                       Pump 2 *or* pump 3 *or* pump 4 runs
                             and
                       Pump 1 *or* pump 3 *or* pump 4 runs
                             and
                       Pump 1 *or* pump 2 *or* pump 4 runs

Failure logic → Success logic when and → or, or → and, and the basic events fail → Basic events succeed.

This equivalence is not immediately obvious, but it can be proven using a branch of mathematics called Boolean algebra, which is specifically designed to deal with combinatorial logic. Boolean algebra is the real key to Fault Tree Analysis.

Figure 9-5 shows the results of applying the same three-step procedure to our logic device example, where success (producing a system output signal) is defined as the receipt of two out of three input signals.

Defining success is critical to the entire process of developing system failure logic. Where do we look for help in defining success criteria when

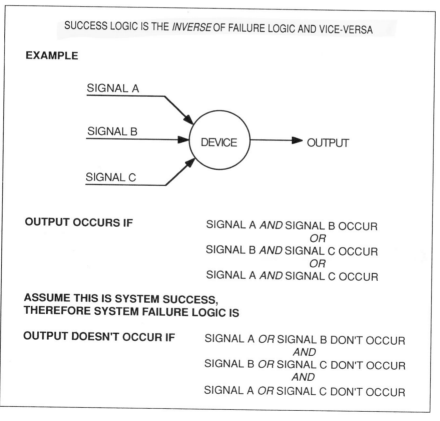

Figure 9-5 Success Logic versus Failure Logic.

we are dealing with complex chemical or process plant systems instead of the simple examples discussed so far? The following is a partial list of sources that can help identify success criteria for a particular plant system or process:

- System design criteria
- Process flow diagrams (PFDs)
- Piping and instrumentation diagrams (P&IDs)
- Other design and operation documents
- Specific calculations.

If the success criteria cannot be extracted from existing documents, it may be necessary to perform specific calculations of system function and capability, to define what *success* means. For example, if system design documentation is incomplete, or if it contains conflicting data (as might be the case for an old system), new calculations might be required. You might need to define the minimum acceptable cooling flow required to maintain a particular reactor temperature. This could then be translated into success criteria in terms of required equipment operating characteristics.

GRAPHICAL SYMBOLS FOR LOGIC MODELING

Describing system success and failure in words works well for very simple systems, but for any realistic system it is better to describe the logic graphically in a fault tree diagram. Typical symbols used in constructing these diagrams are shown in Figure 9-6. A more extensive set of symbols is given in the *Fault Tree Handbook*[1], but those shown in Figure 9-6 are sufficient for all but the most complicated problems you will encounter.

The symbols are grouped into three categories: events, logic gates, and transfer symbols. The *top event* and *intermediate events* are generally shown as rectangles containing a description of the event inside the rectangle, and have other primary or intermediate events which must first occur to make these events possible.

Primary or *basic events* describe the initial faults and require no further development. These events are shown either as circles or as rectangles with small circles attached to the bottom. *Undeveloped events* are treated as *basic events,* but they are actually *intermediate* events that have not been developed further. The reason they are not developed further is either that the information is insufficient or that the consequences are insignificant. *Undeveloped events* are shown either as diamonds or as rectangles with small diamonds attached to the bottom. The *basic events* and *undeveloped events* are the items assigned failure probabilities when the system failure rate is quantified.

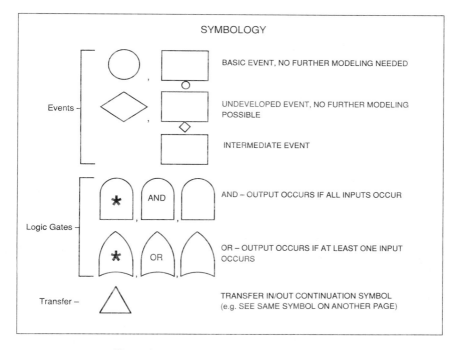

SYMBOLOGY

Events

BASIC EVENT, NO FURTHER MODELING NEEDED

UNDEVELOPED EVENT, NO FURTHER MODELING POSSIBLE

INTERMEDIATE EVENT

Logic Gates

AND – OUTPUT OCCURS IF ALL INPUTS OCCUR

OR – OUTPUT OCCURS IF AT LEAST ONE INPUT OCCURS

Transfer –

TRANSFER IN/OUT CONTINUATION SYMBOL (e.g. SEE SAME SYMBOL ON ANOTHER PAGE)

Figure 9-6   Fault tree construction: Symbology.

The *logic gates* are the symbols used to represent the "and" and "or" logic operators. These logic gates connect the basic and intermediate events to the top event. They can have two or more inputs and provide a single output. Logic gates are shown in several ways. The "and" gate, for example, can be shown with an asterisk or the word *and* inside the symbol, or the symbol can be blank. For an output to occur, *all* the inputs of the "and" gate must be true.

Similarly, the "or" gate can be shown with a plus sign, the word *or,* or nothing inside the symbol. For an output to occur, an "or" gate needs *at least one* of its inputs to occur.

The *transfer* or *continuation* symbol is a small triangle. This symbol is used to connect one part of the tree to another, either for drawing convenience (that is, to avoid the necessity of drawing the same logic over and over again), or to connect parts of the tree if multiple sheets are used to draw it. Typically a letter is placed inside the triangle for transfers from one place to another on the same sheet, and a number is placed inside the triangle when the transfer is from one sheet to another.

If we were constructing a fault tree and got to a point where we placed a rectangle representing an intermediate event near the bottom of the page, if there was no room to continue the development below the rectangle we would put a small triangle, with the number 1 inside it, below the rectangle. On the top of another page, we would then start with a triangle with the number 1 inside. For readability we might repeat the intermediate event rectangle with its description inside, and then proceed with the continuation of the fault tree development.

Now we will use these symbols to create a simple fault tree describing how the electrical circuit shown in Figure 9-7 can fail. The circuit contains a battery, a switch, and two light bulbs in parallel. The light from a single operating light bulb is sufficient for system success. To keep the problem simple, we will ignore the possibility of broken wires or bad connections at the wire terminals.

As we construct the fault tree, the description of the events should go from general categories of failures to more specific categories, and then to specific basic events. Each step is more and more detailed as to the possible causes of the failure.

First we define general system failure modes: no power, and light bulb problems even if power were available. Next we look for specific mechanisms for the above failure modes. These failure mechanisms include switch failure, a dead battery, a switch not properly closed (because of operator error), burned-out bulbs, or loose bulbs.

Figure 9-8 shows the lighting circuit on the left and the simple fault tree

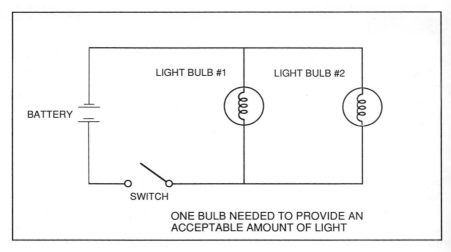

Figure 9-7   Fault tree construction: Simple lighting circuit.

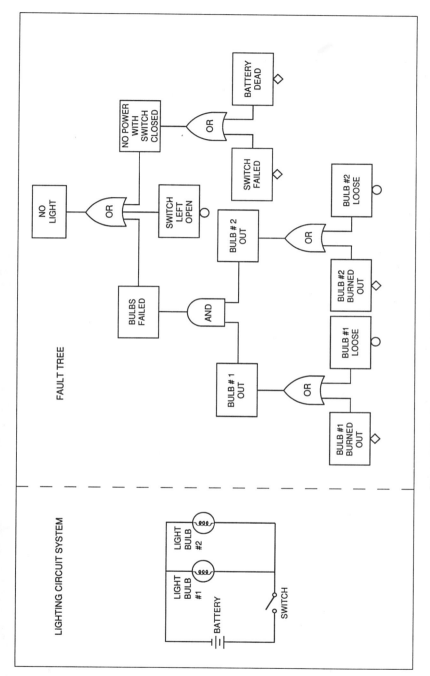

Figure 9-8   Fault tree construction: Single system and fault tree.

137

on the right. The top event in this case is "no light." The next level of intermediate events is connected to the top event by an "or" gate. At this level, we could have shown just two intermediate events, representing the two identified failure modes of "no power" and "light bulb problems with power available"; but the analyst in this case chose to show three events, identifying "switch left open" as a significant way of having no power.

Placing "switch left open" as an intermediate event below an event labeled "no power" and connecting it to that event with an "or" GATE, would be logically equivalent to the way "switch left open" is shown in Figure 9-8. In either case, an occurrence of the event "switch left open" results directly in the top event occurring.

Since it is connected to the top event via "or" gates in either arrangement, no other event has to occur to reach the top of the tree and cause system failure. As long as only "or" gates connect the lower events to the top event, it doesn't matter how far down or how far up the tree they are shown; they are all logically equivalent. However, there are reasons for arranging events in a hierarchy of levels, rather than showing all basic events that connect through "or" gates to the top event at one level.

Note that the branches of the fault tree describing "bulb #1 out" and "bulb #2 out" are identical. Taking advantage of such symmetry in fault tree construction saves a significant amount of time if there are several trains of similar equipment or multiple occurrences of a complex piece of equipment. In these cases, the logic description of one unit is developed once and then copied to other parts of the fault tree, with only minor editing changes to the text identifying particular equipment numbers.

"Bulb loose" and "switch left open" are considered *basic events*. No further breakdown of these events is considered possible, so they are shown with the small circle below the rectangular symbol on the diagram. "Battery dead," "bulb burned out," and "switch failed" are shown as *undeveloped events* with small diamonds under the rectangular boxes.

Although the analyst felt that these events could be broken down further, with several possible causes for each of them (for example, "switch failed" could include a broken part, misalignment, corroded contacts, and other problems), the development was stopped. It could have been stopped at this level because there was no detailed information on the frequency of the different types of component failures, only data on the failure of the part as a whole. The analyst could also have stopped at this level of detail because further detail adds little to an understanding of how the system can fail.

GUIDELINES FOR FAULT TREE CONSTRUCTION

Before discussing the guidelines for fault tree construction, we need to define a few basic terms and make some distinctions between some words that we tend to use interchangeably.

Failure. This is a specific term used to identify the failure of a component to operate properly, or the failure of an operator to perform a certain task.

Fault. This is a more general term than failure and can include the proper operation of an item at an inopportune time as well as the failure of an item to operate properly. An example is the premature closing of a valve by an operator. The valve did not fail, but this event is still called a fault, because it could have serious consequences to the system in question. All failures are faults, but not all faults consist of failures.

Fault category. A component fault can be considered a primary, secondary, or command fault. A *primary fault* is one that occurs within a component's design capabilities—for example, failure at 125 psig of a 150 psig design pressure vessel. A *secondary fault* is failure of the same pressure vessel, but at a pressure in excess of its design pressure. A *command fault* relates to the proper operation of a component at the wrong place or time.

Failure effect. This is the effect of a component failure on a system. A failure effect from our light circuit example is "no power from the battery."

Failure mode. This is an aspect of a component failure. The failure modes corresponding to the "no power from battery" failure effect could be "no electrolyte" or "broken negative terminal."

Failure mechanism. This describes specific ways in which a given failure mode can occur. For example, the failure mechanisms for the failure modes cited above could be "cracked battery case" or "battery sustained mechanical impact."

In this context, failure mechanisms lead to failure modes, which exhibit failure effects on a system.

Active failure. This is failure of a dynamic component that must move or change its state to perform its function. Failure of a motor-operated valve to close on receipt of a signal to close is an active failure.

Passive failure. This is failure of a component that is basically static, such as a pipe, wire, or support beam.

The distinction between active and passive failures becomes more evident when failure rate data are applied to quantify system failure probabilities by assigning failure rates to the individual basic failures contained in a fault tree. Typical failure rates for active components can be several orders of magnitude higher than those for passive components[3,4].

Choosing the Top Event. When constructing fault trees, we generally work from top to bottom. We start with a top event at a system level that usually represents a usable, physical "output" of a system, the lack of which constitutes a system level fault.

Where do we find the top events? Except for the obvious choices, they are typically defined by one of the hazard identification techniques discussed earlier in this book, such as a HAZOP or safety audit. Sometimes the top event of the current fault tree is defined because it is needed as an input to another fault tree or an event tree.

The following general guidelines for fault tree development are derived from the *Fault Tree Handbook*[1]:

Fault Identification—The "Be Precise Rule". Write the fault in the box, being precise as to what fails and when it fails. If you need to abbreviate to make things fit, abbreviate words, but be careful. If you need more room to adequately describe the failure, make bigger boxes!

Immediate Causes—The "Think Small Rule". As you develop the fault tree, try to keep the "steps" from one conceptual level to the next "small," so you gradually progress to lesser levels of abstraction until you finally arrive at the basic event level. Typical levels of abstraction go from the system level, to the subsystem level, to the component level, and finally down to the subcomponent level. We do not necessarily carry fault tree development down to the subcomponent level all the time; in fact, we often stop at the major component level. This gradual development makes it easier to discuss, criticize, and correct fault tree development as it progresses.

A fault tree that jumps immediately from the top event to a long list of basic events connected by "or" gates is not very instructive in terms of revealing how the list of basic events was logically arrived at. By working down the tree from failure modes, to failure mechanisms, to basic component failures, the basic events are grouped into logical categories that can be reviewed for completeness and correctness of the logic involved.

Rule for Abbreviations. Often lengthy descriptions are required to define a fault inside one of the event symbols, and the analyst is forced to abbreviate to fit the description into the box. It is fine to abbreviate, but care should be taken in how abbreviations are chosen, so that no key ideas are lost.

The "No Miracles Rule". If the proper functioning of a particular component, such as a pump or valve, can cause a fault to propagate further along within a system (where failure of the component would prevent the fault from propagating), you should assume that the component performs its function normally. For example, assume that pump A is tripped automatically upon certain plant parameters, and is interlocked to prevent restart until the trip conditions are cleared; pump A could help terminate the event sequence if it was running, but you should not assume that the automatic trip or interlocks "miraculously" fail, allowing manual restart of the pump and termination of the event.

No "Gate-to-Gate" Rule. As you progress down the fault tree, "and" and "or" gates are used to connect several intermediate or basic events to an intermediate or top event above them. Gates should not be directly connected to other gates but should always appear between event rectangles. This is shown in Figure 9-9.

"Complete the Gate" Rule. This rule is intended to force you to develop the tree uniformly across a given level (going horizontally across the fault tree) before you start going into great detail down any one branch (vertically). This helps to ensure that time is not wasted pursuing a particular path in great detail before the overall picture is laid out. Following this rule also helps ensure a more methodical and systematic development.

Figure 9-10 shows a typical example of a simple process system fault tree. This is a simple chemical supply system with a holding tank and two 100-percent-capacity supply pumps in parallel. The top event has been defined as "no flow in outlet line," and success has been defined as the flow from one pump.

All of the major components have been assigned identification numbers, and these are used in the fault tree event boxes to avoid confusion when referring to specific valves and pumps, and to keep the word descriptions brief. The basic event circles and the undeveloped event diamonds have been left off this diagram because of its small size. There are a number of potential common cause failures, such as loss of the power supply, which can fail both parallel pumping paths at the same time.

How detailed should you be in your fault tree modeling? The answer depends on what you are looking for in the final results. You can stop the analysis at a system level, a major component level, or the subcomponent level. The following are some examples of the various levels of analysis details in terms of the way the basic failures are described.

Examples:

1. Pump A fails to run
2. Circuit breaker 50 fails to close
3. Check valve 21C fails to seat
4. Lower motor bearing on pump A fails
5. Shunt coil on circuit breaker 50 open

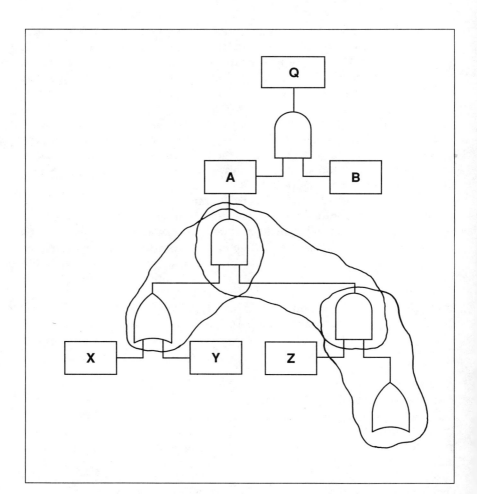

Figure 9-9   Fault tree construction: No gate-to-gate rule.

6. Relay K30 for CB50 open
7. Contact K301 stuck shut

- Examples 1 through 3 are considered component-level basic events.
- Examples 4 through 7 are considered subcomponent-level basic events.
- Most FTA is component-level.

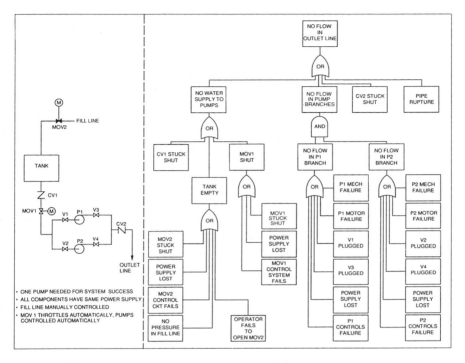

Figure 9-10   Example fault tree: Single pump system and fault tree.

Typical analysis stops at the major component level, but depending on the criticality of the system in question, it can go down to the contact and relay level of control subcomponents if necessary. In general, it is a good idea to carry the analysis at least down to the level where common cause failures can be identified.

BOOLEAN ALGEBRA

As discussed previously, Boolean algebra is the mathematics of combinatorial logic. It expresses in mathematical symbols and equations the logic relationships that we have expressed so far in word descriptions, and graphically as fault trees. In fact, these logic descriptions, and the fault trees themselves, are completely equivalent to a single, long Boolean algebra equation.

Fault Tree Analysis is not just the systematic examination of system failure logic and the construction of fault trees. It includes the mathematical manipulation of the resulting equivalent Boolean algebra equation, to find all of the combinations of equipment failures that can get you to the top event. This list of the combinations of equipment failures (called *cut sets*) is one of the useful outputs of FTA. Luckily, computer codes have been developed that do the hard work of manipulating the Boolean equations. However, to get a better understanding of fault trees and fault tree construction, it helps to understand a little about Boolean algebra.

Boolean algebra concerns itself primarily with variables that can have one of two values: "true" or "false"; "on" or "off"; "1" or "0"; "succeed" or "fail"; "event happens" or "event doesn't happen"; and so forth. It is the mathematics that describes how events are combined and evaluated.

Table 9-1 shows the pure mathematical and engineering versions of the symbols used in Boolean algebra. The concepts represented by these symbols are often depicted graphically, using Venn diagrams, to show the union and intersection of "sets" represented as circles. In Table 9-1 the symbol that looks like a capital letter U represents the Boolean operation of taking the union of two sets. The engineering representation is a plus sign (+), and it represents the "or" gate. Care must be taken not to confuse this with the mathematical operation of addition. They are two very different operations, as we shall soon see.

The next symbol, an upside-down capital letter U, represents the Boolean operation of taking the intersection of two sets. The equivalent engineering representation can be shown three different ways: as an as-

Table 9-1 Boolean Algebra Symbols

Mathematical Symbol	Engineering Symbol	Definition
$\cup$	+	"Or" logic
$\cap$	*, •, NONE (AB)	"And" logic
Ω	1	Universal Set
$\emptyset$	0	Null Set

Note: "Not" operators: $\overline{A}$, A'.

These are Boolean operators, not arithmetic operators (e.g., $A + A \neq 2A$, $A * A \neq A^2$).

Fault tree equivalence:

$+ =$ [OR gate symbol]

$NONE\ (AB),\ *,\ • =$ [AND gate symbol]

terisk (*), as a dot (·), or as no symbol at all between two variables. Again this looks deceptively like the mathematical operation of multiplication, but it definitely is not the same; it represents the Boolean "and" gate.

A capital Greek letter omega (Ω) represents the universal set, or the number 1 in engineering notation, and the Greek letter phi (ϕ) represents the null set, or the number 0 in engineering notation. The "Not" operator is shown as a horizontal bar above the letter representing the Boolean variable. Table 9-1 also shows the fault tree graphic symbols for the "and" gate and the "or" gate.

Figure 9-11 shows a simple fault tree and the equivalent Boolean equation it represents. The top event, Z, is arrived at through an "or" gate from any of the three intermediate events Z1, Z2, or Z3. This is expressed as the Boolean equation

$$Z = Z1 + Z2 + Z3$$

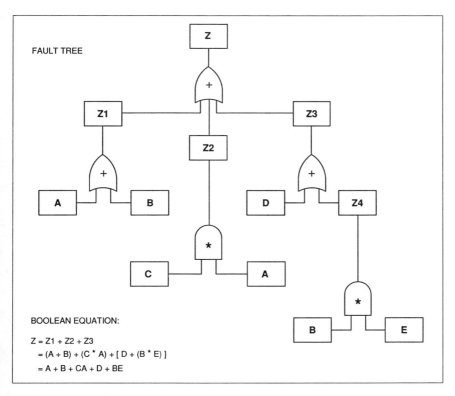

Figure 9-11   Boolean algebra: Fault tree equivalence.

We can now translate each of the other logic gates on the branches of the tree into equations as follows:

$$Z1 = A + B$$
$$Z2 = C * A$$
$$Z3 = D + Z4$$
$$Z4 = B * E$$

Each logic gate in the fault tree is its own small Boolean equation. By substituting variables back into the original equation, the whole fault tree can be represented as a single Boolean equation

$$Z = A + B + CA + D + BE$$

The laws of Boolean algebra help us simplify this equation, which can become very complex for large systems. For real plant systems, the size and complexity of the fault tree and its equivalent Boolean equation are virtually impossible to deal with by hand, and computer programs are almost always used to either draw the trees, or to do the Boolean algebra manipulations.

Knowing a little about Boolean algebra will help you understand how these computer codes work and will also let you analyze simple fault trees without a computer. The following are samples of rearrangement laws that help us simplify the terms in the combined fault tree equation:

$$A * B = B * A$$
$$A + B = B + A$$
$$A * (B * C) = (A * B) * C$$
$$A + (B + C) = (A + B) + C$$
$$A * (B + C) = (A * B) + (A * C)$$
$$A + (B * C) = (A + B) * (A + C)$$
$$(A + B) * (C + D) = (A * C) + (A * D) + (B * C) + (B * D)$$
$$= AC + AD + BC + BD$$

These are examples of the cancellation/absorption laws:

$$A * A = A$$
$$A + A = A$$
$$A + (A * B) = A$$
$$A * (A + B) = A$$

It is in these cancellation laws that the differences between Boolean algebra operations and the more familiar mathematical operations become readily apparent. You might interpret A * A to be A squared, but you must read the equation as A "and" A is equal to A. Similarly you might think that A + A = 2A, but this equation is read as A "or" A = A. The last two equations also give very different results in Boolean algebra than they would in normal algebra.

Figure 9-12 shows a simple Venn diagram proof of the absorption laws.

The following are some of the complementation laws of Boolean algebra:

$$A * \overline{A} = 0$$
$$A + \overline{A} = 1$$
$$\overline{(\overline{A})} = A$$
$$\overline{(A * B)} = \overline{A} + \overline{B}$$
$$\overline{(A + B)} = \overline{A} * \overline{B}$$
$$A + (\overline{A} * B) = A + B$$
$$\overline{A} * (B + A) = \overline{(A + B)} = \overline{A} * \overline{B}$$

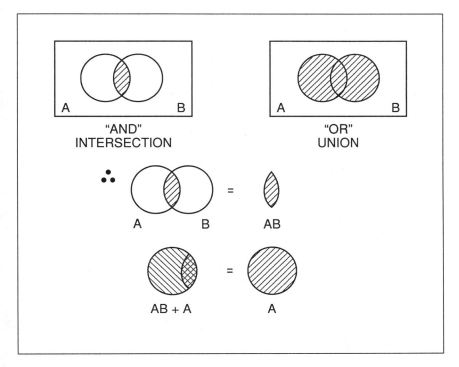

Figure 9-12 Boolean algebra: Absorption laws—proof. A + A* B = A → A⁺ AB = A: Venn diagrams for "And" (*), "Or" (+).

In these equations, A "and" NOT A equals the null set, and A "or" NOT A equals the universal set. Note also that a double inverse cancels itself.

Following is the application of the rearrangement and absorption laws to the fault tree equation derived from Figure 9-11.

$$Z = (A + B) + (C * A) + D + (B * E)$$
$$= A + B + (C * A) + D + (B * E)$$
$$= B + A + (C * A) + D + (B * E)$$
$$= \underbrace{A + (C * A)} + \underbrace{D} + \underbrace{B + (B * E)}$$

$$= A + D + B$$

After all the rearranging is done, note how much simpler the final equation looks than the one we started with:

$$Z = A + D + B$$

This final, reduced equation has a short list of basic events that can lead to the top event of system failure. The process of simplifying the fault tree equation to its minimum number of terms is called fault tree reduction. The minimized list of combinations of events that can lead to system failure is a list of minimal cut sets. If a cut set contains a single basic event, it is a cut set of one (and it represents a possible single failure). You can have cut sets of two, three, or more components, where each cut set represents a possible combination of equipment or operator failures that can lead to the top event. The following helps to explain the definition and shows two examples of minimal cut sets:

$$Z = (A + B) + (CA) + (D + BE) \qquad \text{Eq. 9-1}$$
$$= A + D + B$$
$$\{A\}, \{D\}, \{B\} = \text{cut sets for equation 9-1}$$

They are "minimal" because no more Boolean operations can be performed to reduce the equation further:

$$Z = [A + CD] * [\overline{B} + \overline{E}] * [\overline{B} + \overline{A}] \qquad \text{Eq. 9-2}$$
$$= A\overline{B} + CD\overline{B} + ACD\overline{E}$$
$$\{A\overline{B}\}, \{CD\overline{B}\}, \{ACD\overline{E}\} = \text{cut sets for equation 9-2}$$

They are "sets" because they are independent pathways leading to the top event of the tree.

FAULT TREE REDUCTION AND FAULT TREE EQUIVALENCE

Fault tree reduction is the application of Boolean algebra to the fault tree to reduce it to a list of minimal cut sets. Two fault trees can look completely different, but as long as the basic logic is correctly coded into the structure of the trees, they can still be equivalent in information content and results. The two trees are considered equivalent if the reduction process produces the same minimal cut sets.

The two fault trees shown in Figures 9-13 and 9-14 look very different, but they are completely equivalent. It is very fortunate that fault trees have this property of equivalence, since in practice no two engineers will draw

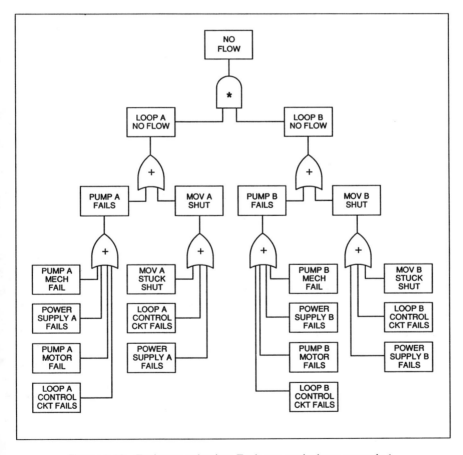

Figure 9-13   Fault tree reduction: Fault tree equivalence example 1.

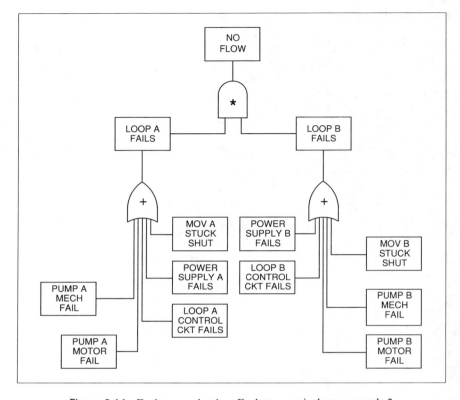

Figure 9-14   Fault tree reduction: Fault tree equivalence example 2.

identical fault trees given the same system or set of equipment. As long as the analysts are systematic about the construction process and they understand the system and how it works, most of the variations in the order of the gates and the appearance of the fault trees wash away in the reduction process to leave equivalent results.

As discussed earlier, in most cases of system level fault trees that represent real world processes, computer programs do the hard work of fault tree reduction and generating the list of minimal cut sets. If these programs are given quantitative failure rate data for each of the basic events, then they can also output the failure probabilities of each of the cut sets and the probability of the top event. Chapters 13 and 14 discuss how to find or develop both the human reliability and equipment failure rate data needed as input to the computer programs.

COMMON CAUSE FAILURES

To reduce failure rates and improve system reliability, good engineering practice includes designing redundancy and diversity into critical plant systems. However, even the best design concepts can be foiled by single events with the potential to fail multiple components. The following is a list of some potential common cause failure categories:

Shared Equipment
Electrical Power
Control Circuits   } Support
Cooling Water   Systems
Pneumatic Supplies
Operator
Common Location
Common Manufactures
Common Maintenance Crew

Figures 9-15 through 9-18 show how shared components and various support systems can be arranged so as to be susceptible to common cause failures. These arrangements are not very unusual, and depending on the

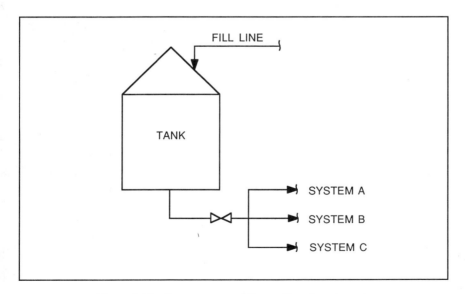

Figure 9-15   Common cause failures: Shared equipment.

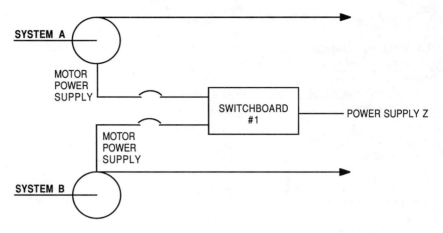

Figure 9-16   Common cause failures: Support systems (power).

reliability of the equipment used and the hazard potential of a system failure, they may or may not be acceptable from a risk or reliability standpoint.

The larger and more complex a system is, the harder it is to spot all the potential common cause failures. Figures 9-19 and 9-20 show the fault tree developed from the system shown in Figure 9-18. Care has been taken to use transfer symbols to properly show the two potential common cause failures of the shared tank, and the common reliance on a single level switch for the control signal to the pumps. In this case, two redundant power supplies and two redundant pumps in each system can be foiled by a

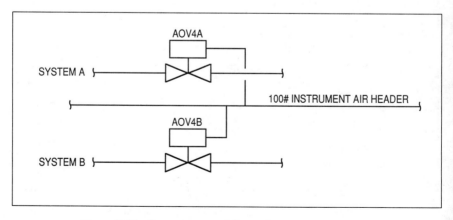

Figure 9-17   Common cause failures: Support systems (air).

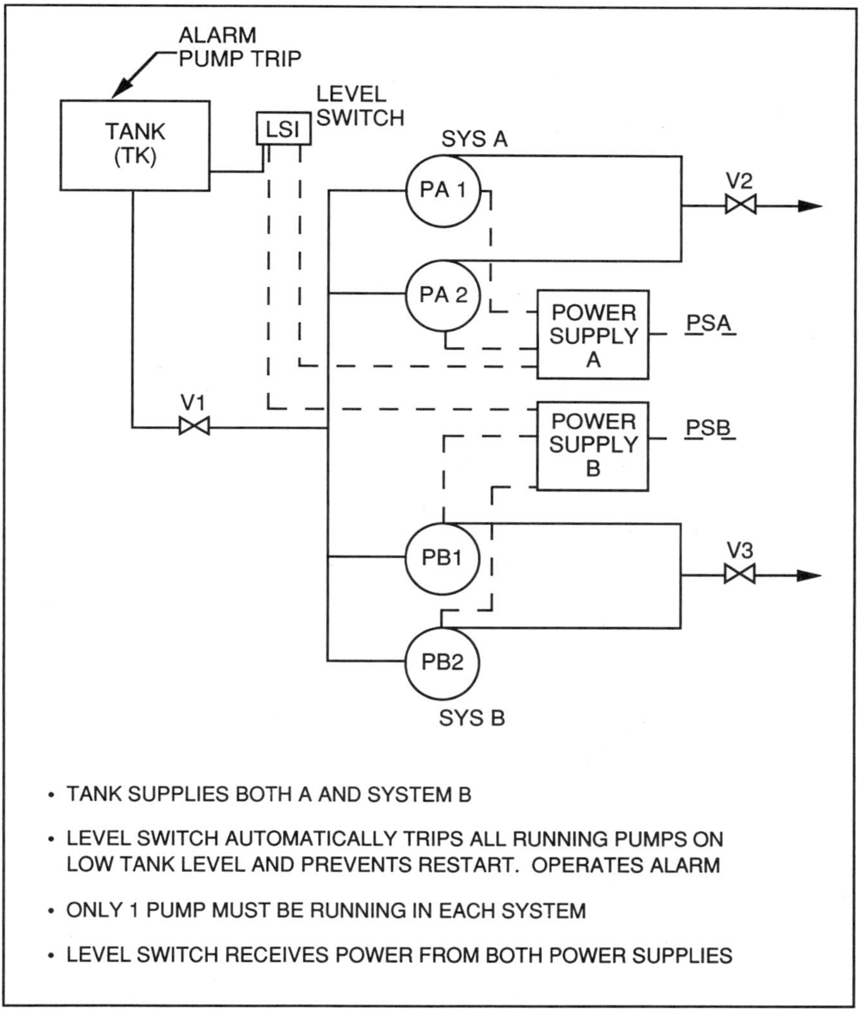

Figure 9-18   Common cause failures: Support systems (controls).

single level switch failure or any one of several single failures relating to the common tank.

The following three steps should be followed in the construction of fault trees to help ensure that all potential common cause failures are identified:

- Devise a coding system for support system components/loops, etc. to ensure consistent basic event labeling.

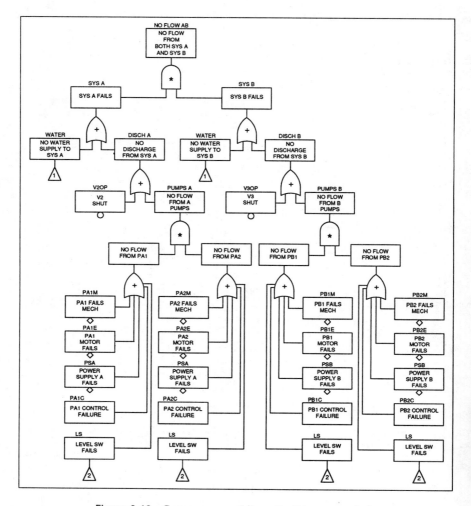

Figure 9-19   Common cause failure: Fault tree example 1.

- Ensure modeling within a system is carried out to a level where all of the support system common points will show up.
- Use transfers to continue modeling of the support system(s) or shared equipment when model will consist of more than one basic event failure.

Carrying the level of detail of the modeling far enough to pick up the various support system functions is essential to picking up potential common cause failures that could otherwise be overlooked.

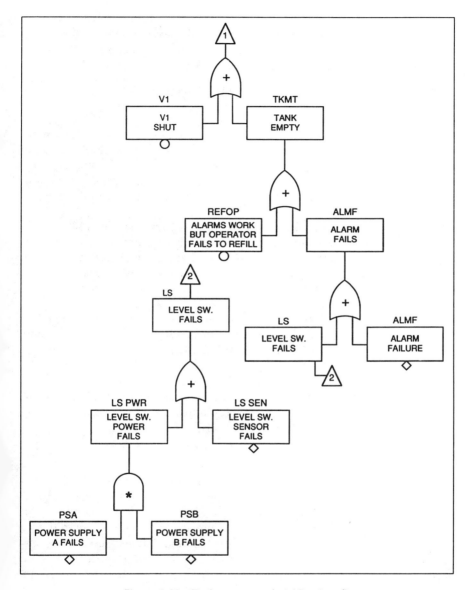

Figure 9-20 Fault tree example 1 (*Continued*).

It is critical to recognize the existence of common cause failures and their potential impact on the calculated system failure rates when you attempt to quantify the fault trees. The common causes must be coded properly into the fault tree structure. It is also important not to attempt to quantify the fault tree itself, but to quantify the reduced tree after the minimal cut sets have been generated. The lists of minimal cut sets, with and without accounting for common cause failures, are always drastically different if common cause failures are present.

EVENT TREES—MODELING MULTIPLE SYSTEMS OR SEQUENCES OF EVENTS

There are many times when you need to model more than a single plant system to find out how a particular hazard can occur or what its potential impact might be. In large-scale plant events or in events involving very hazardous materials, more than one system may be designed to respond. These systems are often designed to operate with a predetermined precedence.

For example, consider a reactor working with a toxic, highly reactive material that is susceptible to a potential runaway reaction. Possible safeguard systems could include a dump system, a chemical additive system, and a flare or burnoff system. To analyze the entire plant response to a runaway reaction accident, a technique for combining models for these three different systems is needed.

One simple way to combine separate models is to make a new top event and place the system fault trees of the other systems as intermediate events below the new top event in a combined fault tree, like the one shown in Figure 9-21.

Suppose that there are several systems playing a role in the response to a given initiating event, *and* there are many combinations of these systems failing and succeeding that produce potentially different events of concern. This kind of situation calls for another tool—the event tree.

Event Trees

Chapters 2 through 8 discussed a host of hazard identification techniques that can be used to identify events of concern and how they can happen. Event trees can be used in either of two ways to account for the interaction of multiple systems or multiple events. They can be used with an initiating event to follow the precursors to an accident and define a series of potential accidents of varying severity, or they can be used to portray possible mitigation and protection systems in which the initiating event is the accident itself.

An example of the first application would be some upset or other initiating event, which if not detected or controlled properly, could lead to a release of a hazardous material. An example of the second application assumes the release of an explosive gas as the initiating event, and then

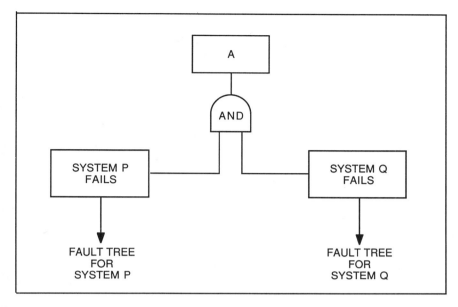

Figure 9-21   Example fault tree: Multiple systems.

considers detection systems, immediate or delayed ignition, and fire protection equipment operation, to determine the spectrum of possible consequence categories (flash fire, unconfined vapor cloud explosion, boiling-liquid-expanding-vapor explosion, or safe dispersal).

Figure 9-22 shows what a typical event tree looks like. In this case, the initiating event is the failure of an automatic process controller. There are three subsequent events that can affect the outcome of the initiating event: a process water system, a glycol cooling system (which depends on the process water system for cooling itself), and an operator who can manually shut down the system.

The result of the various possible sequences of events is a spectrum of possible outcomes, each identified as a specific event sequence (identified by the list of systems that failed in order to reach that particular result). As in the section on Boolean algebra, a system success is designated by a letter with a bar above it, and a system failure is designated by a letter without the bar above it. In Figure 9-22, two sequences lead to minor releases, two lead to moderate releases, and one sequence leads to a major release.

At every step in the sequence, the various systems can either succeed or fail. If we were to quantify the likelihood of either path, the probability of the two possibilities would always add up to 1. With this in mind, we then

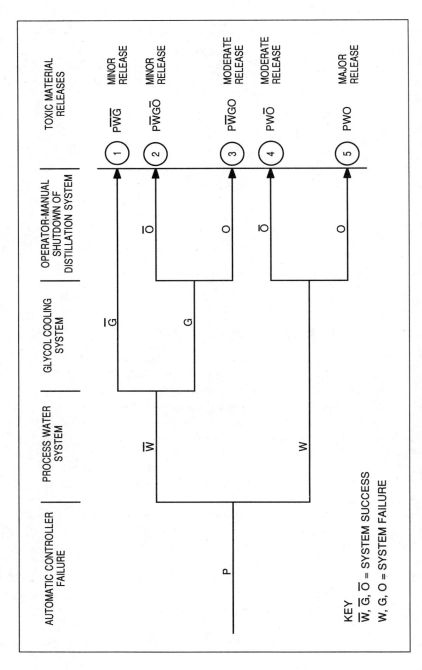

Figure 9-22 Event tree example.

take a second look at Figure 9-22 and wonder: why isn't each branch completely symmetric?

The answer is that this is actually a simplified event tree that takes into account some engineering information about the process water and glycol cooling systems. Its construction acknowledges, for example, that if the process water system fails, there is no cooling water for the glycol cooling system, and it too will fail. Another possible reason that asymmetry could occur is that, given the failure of a particular system early in the sequence, the later systems are inadequate to stop the end result.

Linked Fault Trees and Success Trees

Each branch of the event tree has an identifying sequence of symbols indicating the list of all of the events that either succeeded or failed to get to the end of that branch. Sometimes these sequences are numbered for further discussion, and sometimes they are identified by only including the systems that failed in the symbols that identify the sequence. Figure 9-23 shows the next step in the process. For each sequence of system or event successes and failures, a combined system fault tree is constructed.

If a fault tree was developed for a particular system, and the event sequence includes success of that system, we must first change the system fault tree into a success tree. This conversion is required so that any logical conflicts relating to a component or support system, such as assuming the component fails in one system and succeeds in another, can be detected. This can be done by following the procedure identified earlier, in which we change all of the "and" gates to "or" gates, all of the "or" gates to "and" gates, and all of the failures to successes. This is called creating the "complement" of the fault tree or simply "to complement the fault tree."

Figure 9-24 shows the linked fault and success tree for the third sequence shown in Figure 9-22. The fault tree reduction process will identify potential conflicting assumptions within a particular sequence if they exist. For example, if we did not simplify the event tree and we had a sequence in which the process water system failed and the glycol cooling system succeeded, the conflicting assumption would be identified in the process of reducing the tree to minimal cut sets.

In like manner, if there are any common cause failures that could affect multiple systems, this would be picked up in reducing the combined tree. This process can be defeated if care is not taken in coding the potential common cause failures consistently from system to system.

QUALITATIVE AND QUANTITATIVE ANALYSIS RESULTS

All too often, fault tree analysis is thought of as a method of coming up with a single number, which indicates the probability of a system failure or

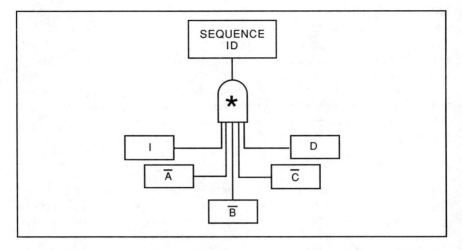

Figure 9-23 Multiple system modeling—Event trees. To analyze each sequence of concern, construct a fault tree for each system or function across the top of the event tree; complement the fault trees for those systems or functions which succeed in the event tree; link the fault/success trees to build a master fault tree that describes each sequence of concern.

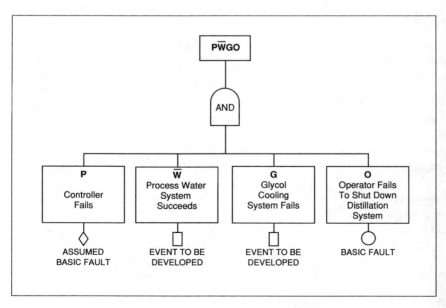

Figure 9-24 Example fault tree: Event sequence P$\overline{\text{W}}$GO.

catastrophic event. While it is possible to produce this kind of result with fault tree analysis, this is by no means the best use, or the only result, of fault tree analysis.

Qualitative Results

Even if we never attempt to apply component failure rates to the basic events in the fault trees and never quantify system failure probabilities, there are many qualitative results of value from fault tree analysis. The first, which is not always so obvious, is the improved understanding of how a particular system works and how it can fail, that we gain by systematically reviewing its operation on a component-by-component basis to construct the fault tree.

After reducing the fault tree to a list of minimal cut sets, we then have a list of *critical components,* which identifies either single components, or combinations of components that can lead to the undesired top event. With this list alone, we are in a position to improve system reliability and reduce the potential frequency of system failure.

When we have quantitative data on component failure rates, it is easier to prioritize the list of critical components, but we can do some prioritizing even without quantitative data. The cut sets of one, or the single failure events, should receive your attention first. After these have been addressed, another approach to prioritizing the list of basic components is to examine the cut sets of two or more components and identify the single components that appear with the greatest frequency in the list. This forms the next list of priority items.

You can now reduce the likelihood of critical component failure or operator error in several ways. Table 9-2 lists a number of possibilities, ranging from relatively simple and inexpensive changes in procedures and inspection frequencies, to equipment replacement, up to system redesign, which is the most expensive option.

Sometimes you cannot address all of the items on the list of critical components, and you need to come up with a better prioritization to minimize the cost and maximize the benefit of the improvements you can make. In this case, you will need to consider some quantitative failure rate information to prioritize the list of cut sets. This quantitative data can still be used in a qualitative manner by considering only the relative failure rates of the cut sets, and not putting a lot of weight on the absolute value of the failure rates.

Prioritizing the list of cut sets with relative failure rate data might uncover some surprises. You could actually have some single failure components that are less critical than some of the two-item cut sets, depending on the individual component reliabilities.

Table 9-2   Risk Reduction Measures for Critical Components

After identifying and prioritizing a list of components critical to system success or failure, these steps can be followed to minimize the likelihood of system failure:
- Increase inspection or preventive maintenance frequency on critical components
- Implement special administrative procedures that take account of the identified system vulnerabilities
- Improve or increase level of operator training relative to critical components or critical operator actions
- Investigate the reliability of critical components, and consider replacing critical equipment with more reliable models if available
- Consider modifying the system design to include more redundancy or diversity

Quantitative Results

Chapters 13 and 14 identify methodology and references that can be used to obtain both equipment failure rate data and predictions of human reliability data, to produce quantitative results from fault trees. An excellent treatment of quantitative risk assessment methodology and directly relevant data for the chemical process industries are presented in two recent volumes from the Center for Chemical Process Safety (CCPS) of the American Institute of Chemical Engineers (AIChE)[5,6].

It is no simple matter to obtain valid equipment failure rate and human reliability data to quantify the fault tree analysis results you develop. Data validity and applicability, statistical analysis, uncertainty analysis, and other issues are dealt with in detail in these other references. We will concentrate here on how to apply the data once the relevant data have been gathered.

The cardinal rules of fault tree quantification are as follows:

- Fault trees are *not* quantified.
- Minimal cut sets are quantified.
- The equation that is the fault tree *must* be in its minimal form (i.e. no more Boolean reduction possible).

Except for a very few exceptions, *you never quantify the fault tree itself.* It is the *minimal cut sets that are quantified,* and these are available for quantification only after the fault tree equation is reduced to its minimal form. You will, more often than not, get incorrect answers if you quantify the tree the way it appears, or the Boolean equation before it is reduced.

The following shows the quantification of a fault tree equation before and after reducing the equation to its minimal form:

$$Z = D + A + (B * C) + A$$

$$\text{Let } A = .1$$
$$B = .02$$
$$C = .03$$
$$D = .01$$

Treating the fault tree operators as Arithmetic instead of Boolean

$$Z = .21$$

When the fault tree is reduced, i.e., the "+" operators are treated as "or" and the "*" operators are treated as "and" first,

$$Z = .11$$

In the reduced form, event A is counted only once (because A "or" A = A), so the results are quite different. It is very tempting, especially with a simple fault tree, to simply show component failure rates right on the fault tree, and then to proceed up the tree adding and multiplying the failure rates to arrive at the failure rate of the top event. However, as the example above shows, this can give incorrect results.

There are only a few exceptional cases in which this procedure would yield correct results. No event in the fault or event tree can be repeated, and all events must be completely independent. In this case, treating the "and" and "or" operators like addition and multiplication operations, and working the numbers progressively up the tree, will result in correct answers.

The quantitative results we get are not limited to determining the frequency of occurrence of the top event of the tree. This number all by itself is seldom very useful. It is only when we compare this frequency of occurrence with that of other events or with that of the same event for cases where we consider variations in the system design, that the result really becomes useful. The quantitative results have their greatest value in comparing the relative risk of different options for risk reduction measures. They give you a solid basis for doing cost-benefit analysis and can provide management with the information required to identify where limited budgets are best spent to get the greatest improvement in plant safety.

As discussed previously, with fault trees that represent real plant systems, computer codes must be used to perform the Boolean algebra reduction to minimal cut sets, and they are also used to perform the quantification.

Table 9-3   Sample Computer Codes for Fault Tree Analysis[a]

Activity	Computer Codes	Availability
Construction of fault trees	Rikke	R. Taylor, Denmark
	CAT	G. Apostolakis
	Fault Propagation	S. Lapp and G. Powers
	IRRAS-PC	EG & G, Idaho
	TREDRA	JBF Associates
	GRAFTER	Westinghouse
	BRAVO	JBF Associates
Qualitative examination	IRRAS-PC	EG & G, Idaho
	CAFTA + PC	Science Applications Int.
	SAICUT	Science Applications Int.
	MOCUS	JBF Associates
	GRAFTER	Westinghouse
	BRAVO	JBF Associates
Quantitative evaluation	IRRAS-PC	EG & G, Idaho
	CAFTA + PC	Science Applications Int.
	SUPERPOCUS	JBF Associates
	GRAFTER	Westinghouse
	BRAVO	JBF Associates
	RISKMAN	Pickard, Lowe, and Garrick

[a] *With permission from the Center for Chemical Process Safety of the American Institute of Chemical Engineers, from Table 3.4 of* Guidelines for Chemical Process Quantitative Risk Analysis.

COMPUTER PROGRAMS FOR FAULT TREE ANALYSIS

There are several areas where computer programs can help us work with fault trees. The computer programs can help draw the trees themselves, perform the fault tree reduction and Boolean algebra, help with the qualitative examination of the cut sets, help perform common cause failure examination, and do the quantitative analysis.

Most of the older computer codes discussed in the *Fault Tree Handbook[1]* were designed to run on mainframe computers and did not have the capability to assist with drawing the trees themselves. These codes were all developed in the late 1960s and early 1970s, but several of them are still being successfully applied today. They are available from the Argonne Code Center at the Argonne National Laboratory.

Computer codes for qualitative analysis included PREP, ELCRAFT, MOCUS, TREEL, ALLCUTS, SETS, and FTAP. Of these, SETS and MOCUS are still in use today. SETS was developed at Sandia National Laboratory and has considerable flexibility, allowing it to deal with complemented events, and special logic gate definitions.

For quantitative analysis, including evaluation of probability distribution functions for failure rates and distributions of repair times, instead of just point values of failure probabilities, KITT, MOCARS, SAMPLE, and FRANTIC were used.

The computer codes COMCAN, BACKFIRE, and SETS allow common cause failure analysis. These codes allow common cause susceptibilities to be associated with each of the *basic events,* and then they examine the cut sets of the fault tree for potential common cause events.

Table 9-3 lists, with permission from *Guidelines for Chemical Process Quantitative Risk Analysis*[5], more current computer codes, many of which run on personal computers. This list also includes programs that help you draw the trees as well as analyze them.

WHEN TO USE AND WHEN NOT TO USE FTA AND ETA

Fault tree and event tree analysis are fairly complex tools to apply, and they take time and a very systematic effort to develop, even with the assistance of computer codes. The techniques discussed in Chapters 2 through 8 all provide means of identifying potential hazards. So when is it appropriate to use fault tree analysis instead of one of the other techniques?

The following is a list of situations in which fault tree or event tree analysis should be used:

- Applicable to a system in which a given undesired event is suspected of being caused by more than one pathway
- Applicable when an undesired event can be stopped by more than one system or function
- Applicable when strong system interactions exist
- Applicable when several support systems exist
- Applicable when the frequency of the undesired event is needed.

Even though a hazard may be identified with one of the other techniques discussed in the earlier chapters, fault tree analysis has a better chance of identifying other possible pathways to the same end result. If there are strong system interactions, or if more than one system comes into play in response to an event, FTA and ETA can handle the analysis. They are also needed if the frequency of occurrence must be determined.

When is it overkill to apply FTA and ETA? The following is a list of some of the situations in which fault tree analysis is not needed or in which there is little to be gained by the extra effort:

- Not applicable when undesired events can be caused only by external events

- Not applicable when undesired event cannot be stopped, slowed, or mitigated by a designed feature of the process
- Generally not applicable when the system or function being considered is not supported by other systems
- Should not be applied to every event at the facility.

Typically, fault tree analysis is applied only to the most critical hazards identified at a plant, after other analyses have been used to identify all of the potential hazards.

SELECTED REFERENCES

1. *Fault Tree Handbook*. January 1981. NUREG-0492. U.S. Nuclear Regulatory Commission. Washington, D.C.
2. Raiffa, H. 1968. *Introductory Lectures on Making Choices under Uncertainty*. Addison-Wesley.
3. *Reactor Safety Study—An Assessment of Accident Risks in U.S. Commercial Nuclear Power Plants*. October 1975. WASH-1400 (NUREG-75/014). U.S. Nuclear Regulatory Commission. Washington, D.C.
4. Lees, F. P. 1980. *Loss Prevention in the Process Industries,* 2 Vols. Butterworths.
5. *Guidelines for Chemical Process Quantitative Risk Analysis*. 1989. Center for Chemical Process Safety of the AIChE. New York.
6. *Guidelines for Process Equipment Reliability Data with Data Tables*. 1989. Center for Chemical Process Safety of the AIChE. New York.

SUGGESTED READINGS

1. NUREG/CR-2300 "PRA Procedures Guide—A Guide to the Performance of Probabilistic Risk Assessments for Nuclear Power Plants", 2 volumes, Published by USNRC, January 1983.
2. NUREG/CR-4350 (Also SAND85-1495) "Probabilistic Risk Assessment Course Documentation", 7 volumes, Available from the National Technical Information Service (NTIS), August 1985.
3. McCormick, Norman J. 1981   "Reliability and Risk Analysis" Academic Press.
4. Watso, J. A. and G. T. Edwards, 1979 "A Study of Common-Mode Failures", R-146, Safety and Reliability Directorate, United Kingdom Atomic Energy Authority, London, England.
5. AIChe/CCPS 1985 "Guidelines for Hazard Evaluation Procedures" Center for Chemical Process Safety, American Institute for Chemical Engineers, New York.
6. NUREG/CR-4780 "Procedures for Treating Common Cause Failures in Safety and Reliability Studies" (also EPRI-NP-5613), published by USNRC.

10

Chemical Plume Dispersion Analysis

Stephen A. Vigeant

Dispersion modeling of postulated accidental releases of hazardous chemicals is an integral part of the overall risk management process. It forms the critical link between the hypothesized equipment failures or release scenarios and the potential consequences suffered by plant personnel and the public. For a given accident scenario involving one or more chemical releases, the potential exists for personal injury or structural damage to the facility due to toxic, flammable, and explosive effects. The dispersion analysis provides the means by which hazardous chemical vapor concentrations can be estimated, both within and beyond plant boundaries, to provide a basis for the quantification of risk.

The physical processes involved in the emission and dispersion of many hazardous chemicals are very complex, and in some cases not very well understood. Much of the complexity of the problem stems from the vast array of possible release and dispersion scenarios that can exist at a given facility. Unlike the dispersion of pollutants emitted from a well-defined and fairly steady-state source, such as a power plant stack, which are still sometimes difficult to accurately model, hazardous chemical releases are typically not well defined and are transient in nature. The release may be instantaneous or continuous from a vessel or pipe involving pressurized gases, refrigerated or pressurized liquids, or liquids at ambient pressure and temperature, resulting in vapor emissions that may or may not be heavier than air. The vapor emissions may be relatively steady-state or may vary with time if from a pressurized vessel or from an evaporating pool of liquid. The releases may involve phase changes and thermodynamic interactions with the environment, with possible rainout of liquid droplets from the plume. Often, plant structures and irregular terrain significantly affect the fate of released chemicals, which further complicates the assessment process.

The breadth of these technical issues involved in hazardous chemical dispersion assessment has necessitated the development of a host of analytical procedures and techniques, some of which can be performed by hand and many that have been computerized. At this writing there are more than 100 mathematical models of varying degrees of sophistication that attempt to address some or most of the physical processes that can potentially be involved in postulated accident scenarios. Many of these models are microcomputer based, while some require mainframe application. Some models are more user friendly than others. The friendly ones may not require a great deal of expertise in their application to a particular problem, while others may require a technical background in chemical processes, thermodynamics, and turbulent diffusion theory.

It is the intent of this chapter to provide an overview of the pertinent issues and theory surrounding the process of analyzing hazardous chemical dispersion for risk assessments, to discuss typical problems and constraints in performing such analyses, and to lend some insight and assistance to the reader in the practical use of available methods and models.

SOURCE DEFINITION

The first step in any dispersion analysis is the characterization of the source of the potential vapor release. The most sophisticated and realistic dispersion models available are of little use unless the specifics of the release can be fairly well defined. The choice of dispersion model used in the analysis is highly dependent on the release scenario.

Releases can originate from any number of plant components, including storage tanks, reactors, and piping. These may be pressurized, refrigerated, or at atmospheric pressure. The chemical released may be a gas stored under pressure or a liquid that is pressurized, refrigerated, or stored at ambient temperature and pressure. The time frame of the release could range anywhere from a few seconds to several hours. The release may occur in a relatively flat and unobstructed area, in an area of irregular terrain, or in the midst of a complex of structures. All of these factors need consideration to model the release properly.

The potential hazardous chemical release pathways are numerous and complex. Accidental releases are generally ill defined, as opposed to routine controlled release of gases from stacks and vents, which are fairly steady-state. Release from storage tanks or vessels can be the result of catastrophic ruptures due to some external event such as a plane crash, cracks resulting from corrosion or fatigue, punctures caused by missiles, or pipeline ruptures. The rate of release in this case is dependent on the specific cause, ranging in duration from a few seconds in the case of a catastrophic rupture, to an hour or more for a puncture or crack.

In a pressurized vessel or tank, a puncture could result in a vapor or liquid jet release. Other potential types of releases include pipe breaks, which could result in either liquid spills or high-velocity liquid or gaseous releases. Uncontrolled releases from flare stacks or vents are also possible as a result of a runaway reaction.

The formation of a toxic or explosive vapor cloud resulting from the various release scenarios is dependent on the nature of the hazardous chemical released and the ambient environmental conditions. For hazardous chemicals stored as liquids under pressure, that have a boiling point below the temperature of the environment, a portion of the release will be instantaneously flashed into a vapor cloud while the remainder of the unflashed portion forms a puddle on the ground. The puddle may spread out unconfined or may be contained in a diked area. The puddle will produce a gradually diminishing plume by vaporization. The vaporization occurs in two ways. First, a rapid vaporization, as heat is initially transferred from both the ground and the air, and then a slower evaporation rate as heat transfer comes primarily from the air (after the ground has cooled). Thus, for this type of stored chemical, the combined effect of the puff release (flashed gas) along with a continuous gaseous plume release (vaporization from the puddle) would need to be evaluated.

For hazardous chemicals stored as liquids that have boiling points above the ambient temperature, the entire release would be evaluated as a continuous gaseous plume due to puddle evaporation. Hazardous chemicals stored as a gas will be released as a vapor cloud, either as a finite puff for instantaneous releases or as a continuous plume for longer term releases.

Where liquids such as anhydrous ammonia are stored under pressure, a catastrophic release may also result in a portion of the liquid being released as an aerosol, in which liquid droplets of various sizes are suspended in the chemical cloud. Unfortunately, it is very difficult to discern what fraction of liquid will be entrained in the cloud for a given release. Studies have shown that this fraction can vary anywhere from 0 to 80 percent, depending on the nature of the release.[1] Generally speaking, the rate of release has much to do with the fraction of liquid in the cloud. Sudden, violent releases of pressurized liquids tend to maximize the amount of liquid thrown into the cloud. On the other hand, a slow release through a pipe leading out of the vapor space in the tank limits the formation of aerosols. A release through an orifice below the liquid level in the tank may result in a significant fraction of liquid being entrained in the released cloud. Some typical release mechanisms are illustrated in Figure 10-1.

Given the complexity of the possible release scenarios, for modeling

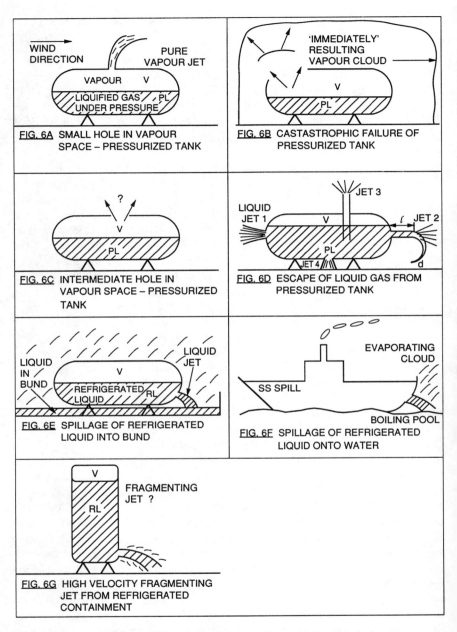

Figure 10-1 Examples of possible release mechanisms. (a) Small hole in vapor space-pressurized tank. (b) Catastrophic failure of pressurized tank. (c) Intermediate hole in vapor space-pressurized tank. (d) Escape of liquefied gas from a pressurized tank. (e) Spillage of refrigerated liquid into bund. (f) Spillage of refrigerated liquid into water. (g) High-velocity fragmenting jet from refrigerated containment. From Fryer and Kaiser, 1979[21].

purposes it is important to characterize the release mechanism in simplified but realistic terms.

SOURCE CHARACTERIZATION

As mentioned in the previous section, determining the source strength as a function of time is a critical element in estimating the vapor dispersion. The accuracy or realism of any dispersion model starts with accurate input data. The determination of the source term itself is dependent on accurate information concerning

- The physical and chemical characteristics of the stored material
- The geometry of the source
- Plant operating procedures
- Spill surface characteristics
- Meteorological data
- Site characteristics such as local topography, buildings, and dikes.

This information is needed to estimate material release rates from the source vessel, flash coefficients, puddle size and spread rate, and liquid pool vaporization or evaporation rate. In many cases, the inputs needed to estimate the source term cannot be accurately determined and must be estimated using sound engineering judgment.

The estimation of material release rates from containment can be a trivial matter or can require fairly complex computer codes, depending on the scenario being evaluated. A scenario involving a catastrophic rupture is easily handled by assuming conservatively that the entire content of the storage tank or vessel is instantaneously released. A hole in a storage tank containing a liquid at atmospheric pressure results in a liquid outflow rate that must be calculated based on the size of the hole, the quantity stored, and the height of liquid above the hole. For a liquid stored under pressure, the storage pressure must also be known to estimate release rate. Releases from vessels or pipes containing gas under pressure generally result in a jet release, if due to a small puncture, and require information on the size of the puncture and gas molecular weight, storage temperature, and gas density. Also, depending on whether the puncture is in the liquid portion of the vessel or in the vapor space above the liquid, the release may consist of both liquid and gaseous phases (two-phase flow), with a release rate somewhere in between that for a pure liquid and a pure vapor.

Catastrophic Ruptures

In a catastrophic tank rupture scenario, the assumption of an instantaneous release of the contents is typically used. Although it is recognized that such an event is not actually instantaneous in that some finite time is required for the tank to be emptied, the assumption is reasonable from a dispersion modeling point of view. The actual time of release is generally negligible compared to sampling times of the resulting concentrations (usually 1 hour). It is also important to consider the fate of such a release since only a portion of the material may become an instantaneous vapor cloud if flashing (rapid vaporization) occurs. The remaining liquid will either spill on the ground, spreading out into a large pool, or be confined within a diked area. In either case, a ground-level plume of vapor will be generated because of vaporization or evaporation over a period of time, depending on the volatility of the liquid. Thus, an instantaneous release may result in a relatively long term source of vapor emissions from evaporating puddles. This type of accident is illustrated in Figure 10-2.

Continuous Liquid Release

In the case of a hole punched in the liquid portion of a pressurized or refrigerated storage vessel (see Figure 10-1), the discharge rate is dependent on the pressure inside the tank, the liquid head, and the size of the puncture. The Bernoulli flow equation is widely used for this purpose and can be expressed as

$$Q = C_d \, A \, \rho_l \left[2 \left(\frac{P_t - P_a)}{\rho_l} + gh \right) \right]^{1/2} \tag{10-1}$$

where

Q = liquid release rate (Kg/sec)
C_d = discharge coefficient (dimensionless)
ρ_l = density of the liquid (Kg/m^3)
A = area of puncture (m^2)
P_t = tank pressure (n/m^2)
P_a = atmospheric pressure (n/m^2)
g = gravitational acceleration (9.8 m/sec^2)
h = liquid head (m)

Values of the discharge coefficient can usually be obtained from standard chemical engineering references such as Perry et al.[2] A typical value of C_d for a pipe break would be 0.8. This method does not account for any

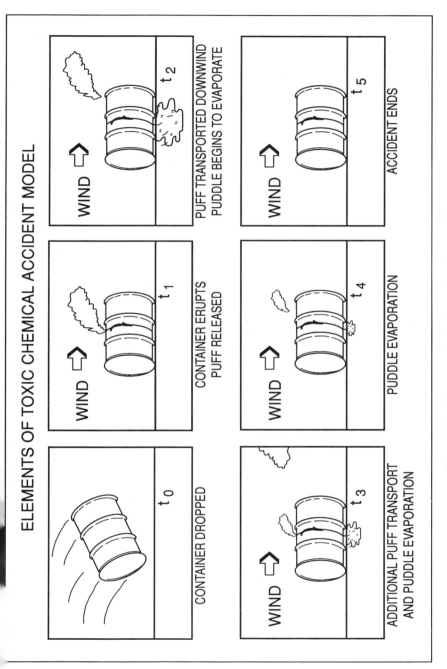

Figure 10-2   Elements of toxic chemical accident model.

173

time dependency of release rate as pressure inside the tank decreases or the liquid head falls. As such, the calculated release rate is an instantaneous value and is conservative if applied over a length of time. Again, if the liquid has a boiling point below ambient temperature, a portion of the liquid will instantly flash to vapor.

Continuous Gaseous Release

Purely gaseous releases from pressurized vessels or pipes generally occur in the form of a jet that can be characterized as critical or subcritical flow. The occurrence of critical or "choked" flow, which has a maximum exit velocity of the speed of sound, is dependent on the ratio of storage pressure to atmosphere pressure. Perry et al.,[2] have expressed the critical flow criterion as

$$P_t/P_a \geq ((\gamma + 1)/2)^{\gamma/(\gamma-1)} \tag{10-2}$$

where P_t and P_a are as defined earlier, and γ is the gamma ratio (heat capacity at constant pressure, C_p, divided by the heat capacity at constant volume, C_v). For many gases, the P_t/P_a ratio that defines critical flow is approximately 2. That is, a storage pressure that is about twice the atmospheric pressure will result in a jet release at the speed of sound. The mass rate of release can then be expressed as

$$Q = C_d \, A \, P_t \, [(M\gamma/RT)(2/\gamma + 1)^{(\gamma+1)/(\gamma-1)}]^{1/2} \tag{10-3}$$

where Q is the gaseous release rate, M is the molecular weight of the gas, R is the universal gas constant, and T is the storage temperature of the gas. This relationship assumes the reversible adiabatic expansion of an ideal gas.

As the storage pressure is relieved following the release, the flow will eventually become subcritical. The expression for critical flow is multiplied by the factor

$$(P_a/P_t)^{1/\gamma}[1 - (P_a/P_t)^{(\gamma-1)/\gamma}]^{1/2} \cdot [(2/\gamma - 1) \left(\frac{\gamma + 1}{2}\right)^{(\gamma+1)/(\gamma-1)}]^{1/2} \tag{10-4}$$

The value of the discharge coefficient C_d is usually less than 1, to account for flow reduction due to viscosity and other discharge effects. A value of 1 for C_d would certainly yield a conservative estimate of outflow. Again, this expression provides an instantaneous flow rate that is actually a function of time. Using initial storage conditions only will yield conservative results.

Two-Phase Releases

Two-phase jet releases consisting of a mixture of gas and liquid can be characterized using empirically derived expressions developed by Fauske[3] or Leung.[4] A good discussion of these methods can be found in Hanna and Drivas.[5]

Flashing Liquids

It was mentioned earlier that liquids with low boiling points may instantaneously flash a portion of the liquid into vapor. This flashed portion can be easily estimated by assuming that the vaporization process is adiabatic. Given this assumption, the heat balance equation to obtain the flashed fraction is simply

$$M_v/M_o = (C_p/H_v)(T_s - T_b) \tag{10-5}$$

where

$\quad M_v$ = mass of vapor due to flashing (Kg)
$\quad M_o$ = total liquid mass (Kg)
$\quad C_p$ = specific heat at constant pressure (J/Kg/°K)
$\quad H_v$ = heat of vaporization (J/Kg)
$\quad T_s$ = storage temperature (°K)
$\quad T_b$ = liquid boiling point (°K)

The vapor cloud due to flashing is normally assumed to be initially at its boiling point, with gradual warming occurring as ambient air is entrained into the cloud. For an instantaneous release, the initial size of the puff caused by flashing can be estimated from the mass released (M_v) and density of the vapor (ρ_v) at its boiling point by assuming a spherical puff as follows:

$$r = [3/4 \ (M_v/\rho_v)/\pi]^{1/3} \tag{10-6}$$

where r is the radius of the puff. For continuous releases of a flashing gas, especially from pipes, it is commonly assumed that all of the liquid flashes to vapor as it is released. In this case, only the liquid release rate need be calculated.

Liquid Pool Evaporation and Vaporization

Another important source of chemical emissions is the liquid pool formed by the spill of chemical with a boiling point above ambient temperature or that portion of a low-boiling-point liquid that does not flash off. In either case, the first step in estimating emission rates due to evaporation or vaporization is to estimate the size of the puddle as it spreads out on the ground. For a relatively small spill, the area of the puddle is easily estimated from the spill quantity and assumed thickness of the puddle. The rate of spread is not important, because it will be of short duration (a few seconds) for most averaging times of interest. For large spills, the rate of spread could have an important effect on estimating initial concentrations, since evaporation or vaporization rate is a linear function of puddle area. By assuming that the initial shape of the spill is in the form of a cylinder with the height equal to the radius of the base, the surface area can be calculated as a function of time using the equation given by Van Ulden,[6] as follows:

$$A = \pi \left[(g \, V_o/\pi)^{1/2} \, 2t + r_o^2 \right] \tag{10-7}$$

where

A = puddle area (m^2)
V_o = volume of the spill (m^3)
t = time from spill (sec)
r_o = initial radius of the spill (m)
g = gravitational acceleration $(9.8 \, m/sec^2)$

The area does not expand indefinitely but reaches a maximum size depending on the vaporization or evaporation rate. Since the contour of the ground cannot normally be well described, the maximum surface area must be calculated by assuming a spill thickness. A typical value of spill thickness assumed for regulatory purposes is 1 centimeter[7,8]. However, if a liquid spill is contained by a dike, the puddle area will be well defined for the purpose of estimating emission rates.

The spill of a liquid with a boiling point above the ambient temperature evaporates by forced convection when exposed to the wind. A commonly used method for calculating evaporation rate can be expressed as

$$\frac{dQ}{dt} = 0.037 \, \frac{D}{L} \, A \, M \, \frac{(PS - PA)}{RT_a} \, R_e^{0.8} S_c^{1/3} \tag{10-8}$$

where

$\dfrac{dQ}{dt}$ = vapor emission rate (Kg/sec)
D = diffusion coefficient (m^2/sec)
L = characteristic length (m)

A   = puddle area (m²)
M = molecular weight (g/mole)
PA = actual vapor pressure (n/m²)
T_a = air temperature (°K)
PS = saturation vapor pressure (n/m²)
R_e = Reynolds number = $LU\rho_a/\mu$ (dimensionless)
S_c = $\mu/D\rho_a$ = Schmidt number (dimensionless)
μ = kinematic viscosity (Kg/m.sec)
U = mean wind speed (m/sec)
ρ_a = density of air (Kg/m³)

The value of the characteristic length (L) can be chosen to be equal to the diameter of the liquid pool. Also, the value of the diffusion coefficient (D) can be obtained from standard chemical reference books. However, if values for D are not readily available, a value of 2×10^{-3}m²/sec would generally yield conservative evaporation rates[8].

In the case of a liquid with a boiling point below that of the environment, the liquid pool vaporizes by absorption of heat from long-wave radiation, short-wave solar radiation, convection of air, and ground conduction (see Figure 10-3). Initially, the conduction of heat from the ground dominates the heat budget and can be estimated according to

$$q_d = K_s \, A \, (T_e - T_b)/(\pi K_s t/\rho_e \, C_{pe})^{1/2} \tag{10-9}$$

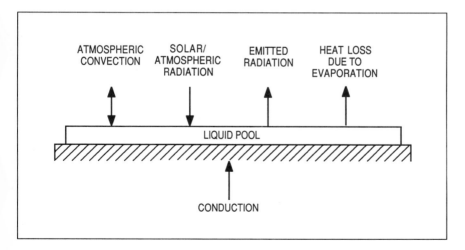

Figure 10-3 The heat budget of an evaporating pool. Shaw and Briscoe, 1978, p. 78[9].

where

q_d = heat transfer rate (watts)
K_s = thermal conductivity of the soil (watt/m/°K)
T_e = soil temperature (°K)
T_b = chemical boiling point (°K)
ρ_e = density of earth's crust (Kg/m^3)
C_{pe} = heat capacity of the earth's crust (J/Kg/°K)

Since soil conditions cannot always be known or well defined, typical values of K_s, ρ_e and C_{pe} can be used such as 1.67 watt/m/°K for K_s, 1,520 Kg/m^3 for ρ_e, and 837 J/Kg/°K for C_{pe}[8].

In the same vein, the heat flux due to solar and atmospheric radiation (q_r) can be conservatively estimated if specific conditions are not known for the analysis at hand. A value corresponding to noontime between June 1 and July 1 and 30°N latitude (1,150 watts/m^2) could be used as a conservative estimate of this parameter[8].

The heat flux due to forced convection (q_c) can be estimated according to

$$q_c = h_c (T_a - T_b) \qquad (10\text{-}10)$$

where h_c is a heat-transfer coefficient (watts/m^2/°K) and T_a is the air temperature. A value of 6.7 watts/m^2/°K corresponding to a wind speed of 1 m/sec could be used for this parameter[10].

The vaporization rate is then the result of the sum of the different heat fluxes as follows:

$$\frac{dq}{dt} = (q_d + q_r + q_c)/Hv \qquad (10\text{-}11)$$

A good discussion of other formulations for evaporation and vaporization calculations, along with comparisons with field tests, can be found in Hanna and Drivas[5].

DISPERSION MODELING

Thus far, the discussion has brought us to the point of beginning to consider the dispersion of a hazardous chemical release, indicating the potential complexities of the subject matter. Now that the nature of the accident and the source term have been explored and dealt with, the final task of analytically addressing dispersion begins.

As in the case of the source term definition, the potential dispersion scenarios are many. Instantaneous and continuous releases are handled with different methodologies, as are neutrally buoyant and heavier-than-air gases and violent or passive releases. However, the dilution or dispersion of a released gas or aerosol can be thought of in the context of three basic mechanisms: mechanical turbulence, turbulence due to buoyancy, and atmospheric turbulent fluctuations. Although atmospheric turbulence is actually comprised of both elements (mechanical and thermally induced), the mechanical turbulence referred to here is that caused by mechanical energy imparted to the release by some sort of violent process, as in the case of a pressurized container rupture. The turbulence due to buoyancy refers to that caused by density differences between the cloud and ambient air. Atmospheric turbulence is that caused by random fluctuations in the wind field caused by both mechanical (wind flowing over rough ground) or thermal (temperature stratification) forces.

The mechanical turbulence caused by violent releases is generally treated through some sort of entrainment algorithm in which entrainment coefficients based on experimental data are used to estimate the proper amounts of heat, mass, and momentum entrained into the cloud upon release. Buoyancy-induced turbulence is generally dealt with through a box or slab model in which concentration of a chemical is assumed uniform and ambient air is drawn in through the edges and top of the box using a variety of entrainment functions. In some cases, a distribution function is assumed for the concentration profile within the box or slab.

Atmospheric turbulence is almost always treated using the well-known Gaussian dispersion model, which assumes a normal distribution of material concentration within a plume or puff using values of standard deviations in the horizontal and vertical based on experimental data[11].

This section addresses the common analytical approaches used in simulating these dispersion mechanisms, including inherent limitations and uncertainties associated with these methods. Some model validation efforts and available experimental data used in this regard will also be examined and discussed, leading to a general understanding of the current state of the art of hazardous chemical dispersion modeling.

Jet Release Dispersion

The beginning of the dispersion process depends on how the hazardous chemical escapes to the environment. As discussed earlier, the escape routes are quite varied. A tank or drum rupture spilling a high-boiling-point liquid on the ground produces a ground level plume because of evaporation. The density of the plume relative to air will determine whether turbulence from buoyancy or atmospheric turbulence dominates the initial dispersion process. The same applies to a vaporizing puddle of a low-boiling-point chemical spill. In this case, the flashed portion of the release may entrain ambient air during adiabatic expansion. A conservative approach would be to assume that the flashed portion is composed of pure chemical with no dilution upon flashing.

In the instance of a pressurized release from a pipeline or vessel (see Figure 10-1), ignoring entrained ambient air upon release is unduly conservative. This initial dilution upon release can vary widely, depending on the exact nature of the release. Observations of either actual accidental releases or controlled experiments indicate initial dilutions of 10- to 100-fold or greater in some cases[1]. In the absence of an analytical estimate of initial dilution, a factor of 10 is a reasonably conservative estimate in most cases of jet releases or violent vessel ruptures.

Analytical approaches to estimating jet release dilution generally consist of integral-type models in which equations for the conservation of physical properties such as mass, momentum, and heat are solved numerically or in a stepwise fashion. Since this set of equations cannot be solved exactly, a parameterization scheme is generally used whereby entrainment coefficients are used to quantify the interaction between the jet properties and environmental properties such as mass and momentum entrainment. These coefficients are generally derived from field or laboratory experiments such as in a wind tunnel. Although this approach introduces a certain degree of error in the analysis, since the actual physical interactions taking place are not accounted for, it does provide a reasonable estimate of bulk properties of the jet. The result is a generally good approximation of the gross features of the jet, such as centerline concentration and jet radius as a function of downwind distance.

This type of jet model can be used to describe the properties of the jet (concentration, temperature, width) in the near field and can be coupled with a Gaussian dispersion model at the point where atmospheric turbulence dominates the dispersion or with a heavy gas dispersion model. This type of approach has been successfully applied in the DEGADIS model, for example[11]. A good discussion of jet modeling techniques can be found in Ooms[12] and Hanna and Drivas[5].

Heavy Gas Dispersion

Density-driven or heavy gas dispersion is a topic that has been of interest and investigated only since the mid-1970s. Previously, only dispersion caused by atmospheric turbulent fluctuation was keenly studied. Hanna and Drivas[5] note that Van Ulden's[6] experiments with dense gas clouds caused many investigators to consider ways other than Gaussian models to account for dispersion in such situations. Indeed, much effort has been put forth in the modeling of heavy gas dispersion since that time, and a great body of literature now exists on the subject. Among the many references available on the subject are Hanna and Drivas[5], McNaughton et al.[13], Eidsvik et al.[14], Woodward[15], Colenbrander[16], Tasker[17], and Spicer et al.[18]

As the name implies, the main driving force in "heavy gas" dispersion is the weight or density of the cloud or plume relative to that of ambient air. The weight of the cloud causes it to slump and spread out in all directions. The weight of the cloud may be due to a high molecular weight or low temperature, as in a refrigerated gas.

The rate of spread of the cloud (dr/dt) is generally expressed as follows:

$$\frac{dr}{dt} = [gh(\rho_c - \rho_a)/\rho_a]^{1/2}$$

where h is the height of the cloud, assuming that it is in the form of a cylinder. To estimate concentrations in the cloud, its dilution with ambient air must also be estimated. This is generally handled using entrainment assumptions and assuming that concentration is uniform within the cloud, the so-called box or slab model approach. A variety of entrainment assumptions are used in models with some accounting for both entrainment through the front and sides of the cloud (edge entrainment) as well as from the top of the cloud (top entrainment). These entrainment assumptions usually take the form of an entrainment coefficient (α) multiplied by the frontal velocity (V_f) of the cloud for edge entrainment (αV_f) and another entrainment coefficient (β) times a velocity (U or V) divided by the Richardson number (RI) for top entrainment. Some models account for either edge or top entrainment, and there are variations on the entrainment assumptions. Again, the entrainment coefficients are experimentally derived values.

It is also important to note that for instantaneous releases, the finite puff is free to spread in all directions, making edge entrainment an important factor. However, for continuous releases, the spread in the downwind and upwind directions is restricted by the continuous release of gas, making top entrainment more important. Thus, continuous plumes tend to spread faster laterally than an instantaneous puff.

Thus far, the discussion on heavy gas dispersion has focused on entrainment of ambient air which dilutes the cloud and reduces its density. However, many times there are also thermodynamic processes at work that have a profound effect on the cloud's temperature and thus its density (see Figure 10-4). As the cloud moves downwind, heat is absorbed from

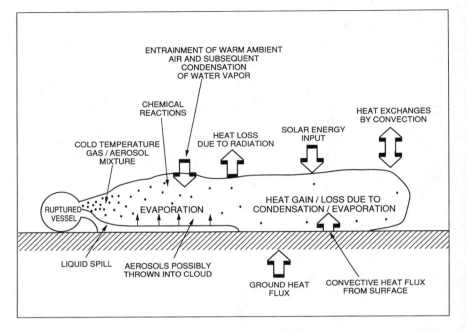

Figure 10-4   Thermodynamic aspects of a typical hazardous material release. From Hanna and Drivas, 1987[5].

the ground as well as from solar radiation, convection from the air and heat released from condensation of chemical or water vapor. Heat is lost from the cloud due to radiation losses and evaporation of any chemical or water droplets in the cloud. Any chemical reactions that may be taking place in the cloud also contribute to the heat budget. In most cases, such reactions are not accounted for in models. There are some models specifically formulated to account for reactions taking place in a hydrogen fluoride cloud, but these tend to be very specialized models. Some models account for thermodynamic processes and others do not. The importance of thermodynamics to the dispersion process depends very much on the chemical released. However, it is usually desirable to account for as many of these processes as possible to be physically realistic while maintaining reasonable economy in the execution of these models.

Neutrally Buoyant Gases

At some point in the process of heavy gas dispersion, density-driven turbulence becomes weak as ambient air is entrained in the cloud and atmospheric turbulence begins to dominate the dispersion process. Various criteria are used to define this point of transition. Some modelers have chosen a critical value of the local Richardson number below which the transition occurs. Several others have chosen a critical value of the cloud frontal velocity (V_f), usually expressed as some fraction or multiple of the friction velocity (U_*). Certainly the choice of a criterion for this shift in the dominant dispersion mechanism will have a direct bearing on the predicted concentrations within a certain range of distance. According to Hanna and Drivas[5], most researchers seem to agree that the value of V_f should approximate U_* at the point of transition.

Once the transition from buoyancy-driven dispersion to atmospheric turbulent diffusion occurs, a virtual point source calculation is typically performed to match the box model with a Gaussian puff or plume model. This technique derives a distance from which a release would be dispersed by atmospheric turbulence to a concentration that equals that predicted by a box model for dense gas dispersion. In this way, the dispersion calculation can continue from the transition point in a standard Gaussian plume or puff manner.

Given the amount of a hazardous substance released as a puff and the emission rate from a spill, the downwind concentration of a neutrally buoyant gas due to dispersion by atmospheric turbulence is calculated by assuming that the material is distributed in the puff or plume in a Gaussian manner (see Figure 10-5). For a puff release with an initial finite volume, the chemical concentration in the puff is described by the following equation[19]:

$$X(x, y, z, h) = \frac{2Q}{(2\pi)^{3/2} \, \sigma_{xI} \, \sigma_{yI} \, \sigma_{zI}} \cdot \exp\left[-1/2\left(\frac{x^2}{\sigma_{xI}^2} + \frac{y^2}{\sigma_{yI}^2}\right)\right]$$

$$\cdot \left\{\exp\left[-1/2\,\frac{(z - h)^2}{\sigma_{zI}^2}\right] + \exp\left[-1/2\,\frac{(z + h)^2}{\sigma_{zI}^2}\right]\right\}$$

The initial volume of the puff is accounted for by adjusting the standard deviations of concentration in the following manner:

$$\sigma_{xI}^2 = \sigma_{xI}^2 + \sigma_o^2$$
$$\sigma_{yI}^2 = \sigma_{yI}^2 + \sigma_o^2$$
$$\sigma_{zI}^2 = \sigma_{zI}^2 + \sigma_o^2$$
$$\sigma_{xI}^2 = \sigma_{yI}^2$$

where

$$\sigma_o = [M_v/(2^{1/2} \cdot \pi^{3/2} \cdot \rho_v)]^{1/3}$$

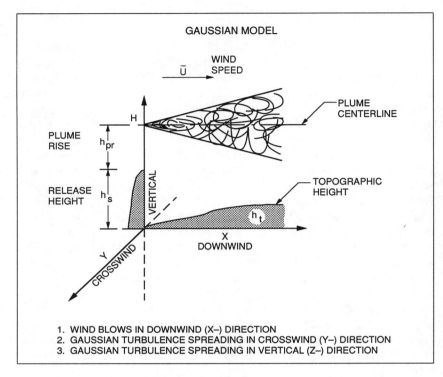

Figure 10-5   Gaussian model.

The value of x in the exponential term is determined by:

$$x = x_o - U t$$

where x_o is the distance between the source of the spill and the receptor.

The diffusion equation for a continuous ground-level plume release is given by[19]:

$$X(x, y, z, h) = \frac{Q'}{2\pi \, U \, \sigma_y \, \sigma_z} \cdot \exp\left[-1/2 \, \frac{(y^2)}{\sigma_y^2}\right]$$

$$\cdot \left\{\exp\left[-1/2 \, \frac{(z - h)^2}{\sigma_z^2}\right] + \exp\left[-1/2 \, \frac{(z + h)^2}{\sigma_z^2}\right]\right\}$$

The finite initial size of the spill is accounted for by replacing σ_y with $(\sigma_y^2 + \sigma_{yo}^2)^{1/2}$, where σ_{yo} is approximated by[19]

$$\sigma_{yo} \approx r^{1/2}/4.3$$

and r is the radius of the spill. The values of the dispersion coefficients (σ_y and σ_z) are taken from the Pasquill-Gifford curves[19].

Model Performance and Uncertainty

Despite all the effort expended on developing analytical techniques to simulate the physical mechanisms responsible for the dispersion of a released chemical, most dispersion models fall far short of the realism ideally desired for consequence and risk assessment purposes. This is due in large part to the inherent complexity and randomness associated with these processes coupled with the numerous uncertainties and lack of accurate and detailed input data that define the release scenario of interest. Even the most sophisticated numerical models that actually attempt to solve the fundamental physical equations that govern atmospheric motion and turbulence are limited by the random nature of turbulent fluctuations and the fact that the equations do not have exact solutions. In addition, such models require input data that are normally not available and require extensive mainframe computing capability that is impractical for almost all applications. Thus, to make the calculations manageable, the physical equations are solved analytically by assuming steady-state conditions and using empirically derived parameters to close the set of equations. This simplified solution, in which experimental data for a particular location or situation are used for general application, leads to inaccuracies that cannot be overcome.

The Gaussian model described in the previous section falls into the category of a simplified solution to the dispersion problem. The values of the dispersion coefficients σ_y and σ_z commonly used are derived from dispersion studies conducted at a specific location under a specific set of conditions and sampling times that are probably not applicable to the analysis being performed. In addition, there are a number of assumptions that should be met for the model to be consistent with the situation being analyzed. These include

- Steady-state condition for wind speed, direction, and atmospheric stability over the period of the simulation
- Spatial uniformity of meteorological parameters over the domain of the simulation

- Terrain features uniform over the domain of the simulation and similar to those present in the development of the dispersion coefficients
- Source emissions constant over the averaging time of the model
- Pollutant mass conserved throughout the dispersion process.

The key point to be made here is that dispersion models contain many simplifications and assumptions that lead to predictions of pollutant concentrations that at best can be considered estimates. Several model evaluation studies have been performed under EPA sponsorship in an attempt to quantify the accuracy of these Gaussian model estimates. They indicate that accuracy is very much dependent on the complexity of terrain and meteorological conditions and on how comparisons between predictions and observations are made. Under fairly homogeneous and steady-state conditions with no exceptional circumstances and with reasonably representative meteorological data of good quality available, accuracy within a factor of 2 is probably realistic[20]. Outside these constraints, accuracies can easily fall into the order of magnitude range (factor of 10).

It is also important to consider what is meant by accuracy. Comparing predicted versus observed 1-hour concentrations on a paired basis (i.e., at the same point in time and space), which is the true measure of model performance, is quite different from comparing maximum predicted versus observed concentrations regardless of where and when they occurred. In many cases, model performance has been shown to be reasonably good when viewing a cumulative frequency distribution of highest observed and predicted concentrations (e.g., see Figure 10-6 taken from McNaughton et al.[13]). However, the scatter of a plot of paired observed versus predicted concentrations shows a somewhat different picture of model performance (see Figure 10-7 from McNaughton et al.[13]). These observations imply that dispersion models are generally better able to predict expected values from a distribution of values rather than a distinct value in response to a specific set of conditions which is normally desired in hazardous chemical modeling.

In the case of hazardous material models, there are additional constraints that add further to uncertainty of predictions, such as

- Heavy gas dispersion
- Non-steady-state releases
- Thermodynamic effects
- Aerosol formation.

The performance of this class of dispersion models has been tested in many different studies using the available experimental data on hazardous

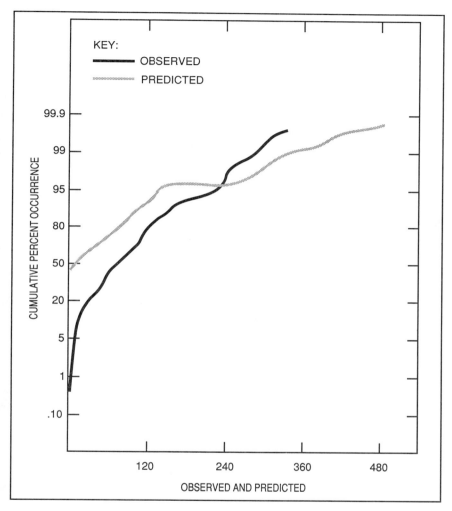

Figure 10-6   Cumulative frequency distributions. From McNaughton et al., 1986[13].

chemical releases. The vast bulk of these experiments deal with heavy gas dispersion and are limited in scale and variety of conditions studied, mainly because of safety and cost considerations. These validation studies are typically limited to relatively simple statistical measures of performance, such as plots of maximum predicted concentration versus distance and ratios of predicted to observed maximum concentrations. The results

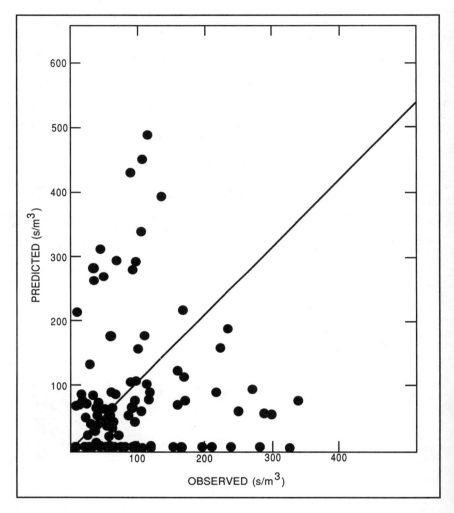

Figure 10-7   Comparison of predicted vs. observed concentrations. From McNaughton et al., 1986[13].

are highly variable, showing good agreement of predictions with observations in some cases (within a factor of 2) and very little agreement in others. More detailed discussions of model evaluation and validation studies can be found in Hanna and Drivas[5], McNaughton et al.[13], and Woodward[15].

SPECIFIC MODELS

As mentioned earlier, several computer models have been developed to address some or all of the aspects of hazardous chemical dispersion analyses. Table 10-1 describes various dispersion models that can be used to assess the impact of accidental releases of hazardous materials.

Hanna and Drivas[5], and McNaughton et al.[13] provide excellent discussions of the available models at the time of their publication. Some models are available for use on personal computers at little or no cost. Many of these models are commercially available, and the developers should be contacted for specific terms of obtaining them.

The EPA dispersion models as well as DEGADIS can be obtained for the cost of the telephone call to an electronic bulletin board service operated by EPA called the Support Center for Regulatory Air Models (SCRAM). The source codes as well as executable code can be downloaded from the SCRAM bulletin board to a PC. The other models can be obtained by making the appropriate arrangements with the developers.

SAMPLE PROBLEMS

This section provides examples of typical hazardous release scenarios and how the analytical techniques discussed in the previous section can be used in assessing potential consequences. Typical assumptions in regard to input data and methods of analysis and application of the results are included.

Ammonia Storage Accident

In this scenario, 30 metric tons of anhydrous ammonia are being stored in a refrigerated tank. The storage pressure is 1 atmosphere and the storage temperature is $-33°C$ or $240°K$. A break develops in a 2-inch line from the tank leading to a liquid outflow onto the ground in an unconfined area. The outflow rate can be estimated from the Bernoulli flow equation (10-1) given the liquid head of 5 meters and the area of the discharge (area $= \pi (0.0508 \text{ m/2})^2 = 2.03 \times 10^{-3} \text{ m}^2$), assuming a typical discharge coefficient for a pipe break of 0.8, as follows:

$$Q(\text{Kg/s}) = 0.8 \times (2.03 \times 10^{-3} \text{ m}^2) \times (674 \text{ Kg/m}^3) \times [2 \times (9.8 \text{ m/s}^2) \times (5 \text{ m})]^{1/2} = 10.8 \text{ Kg/s}$$

Table 10-1   Examples of Available Dispersion Models

Model	Type	Developer	Computer	Cost
INPUFF	Puff/Gaussian/ Variable trajectory	EPA	PC	None
TXDS	Continuous plume/ Gaussian	NJ Dept. of Environmental Protection	PC	None
DEGADIS	Jet release/Heavy gas dispersion	Spicer et al., 1986/EPA, 1989	DEC VAX PC	None
CAMEO II Air Model	Continuous plume/ Gaussian	NOAA—Hazardous Mat. Response Branch	PC/ MacIntosh	Yes
ARCHIE	Toxic cloud/Pool fire explosions/ Fireballs/Flame jets/Source model	EPA/FEMA	PC	None
WHAZAN	Puff/Continuous plume/Gaussian/ Heavy gas/ Source emissions models/Jet releases/ Explosions/ Fireballs/Flame jets	Technica, Int.[20]	PC	Yes
SPILLS	Pool evaporation/ Puff/Continuous plume/Gaussian	Shell Devel. Center	PC	Yes
HASTE	Source emissions/ Puff/Continuous plume/Gaussian/ Heavy gas/Real-time use	ENSR	PC	Yes
CHARM	Source emissions/ Jet/Puff/ Continous plume/Gaussian/ Heavy gas/Real-time use	Radian	PC	Yes
CARE	Source emissions/ Puff/Continous plume/Gaussian/ Heavy gas/Real-time use	Environ. Systems Corp.	PC	Yes

(cont'd)

Table 10-1   (*Continued*)

Model	Type	Developer	Computer	Cost
MIDAS	Source emissions/ Jet/Puff/ Continuous plume/Gaussian/ heavy gas/Real- time use	Pickard, Lowe & Garrick	PC	Yes
PHAST	Source emissions/ Jet/Puff/ Continuous plume/Gaussian/ Heavy gas	Technica, Int.	PC	Yes

where the density ρ_t of liquid anhydrous ammonia is 674 kg/m^3. Since the anhydrous ammonia is refrigerated and the storage pressure is only 1 atmosphere, the only driving force for outflow is the liquid head.

Since the storage temperature is equal to the boiling point of ammonia (240°K), there is no flashing of the liquid that spills on the ground and spreads out into a puddle. A commonly assumed spill thickness is 1 cm. The puddle will spread only as much as the vaporization rate will allow. The vaporization rate will increase with time as heat is absorbed by the pool, mainly from ground conduction initially, but also from solar and atmospheric radiation. It may happen that the ground conduction component of the heat budget will drop off substantially as the ground cools, but the conservative approach would be to ignore this effect. An example of the evaporation rate calculation using equations 10-9 and 10-10 is given in the following equation, assuming an ambient temperature of 90°F (305°K) and t = 10 seconds:

$$q_d/A = (1.67 \text{ watt/m}^{2°}\text{K}) \times (305°\text{K} - 240\text{K})/$$
$$[(\pi \times (1.67 \text{ watt/m}^2 \text{ K}) \times (10 \text{ s})/(1,520 \text{ Kg/m}^3)$$
$$\times (837 \text{ J/Kg}°\text{K})]^{1/2} = 16,904 \text{ watt/m}^2$$
$$q_r \text{ (assumed)} = 1,150 \text{ watt/m}^2$$
$$q_c = (6.7 \text{ watt/m}^2 \text{ K}) \times (305 \text{ K} - 240 \text{ K}) = 435.5 \text{ watt/m}^2$$
$$dq/dt = (16,904 + 1,150 + 435.5 \text{ watt/m}^2)/(1.37 \times 10^6 \text{ J/Kg})$$
$$= 0.0135 \text{ Kg/s/m}^2$$

This calculation illustrates the initial dominance of the ground conduction component of the vaporization process.

The actual evaporation rate at this time (10 s) can then be determined from the puddle area, which is estimated using equation 10-7 as follows:

$$A = \pi \, [((9.8 \text{ m/s}^2) \times (108 \text{ Kg}/674 \text{ Kg/m}^3)/\pi)^{1/2} \times 2(10 \text{ s})] = 44.41 \text{ m}^2$$

In this example, the initial radius r_0 is ignored, since the release area corresponds to only a 2-inch-diameter pipe. This calculation also assumes that the 10-second release period (10.8 Kg/s $\times$ 10 s = 108 Kg) is small enough to be considered instantaneous. Thus, the vaporization rate 10 seconds into the spill is estimated to be

$$dq/dt = (0.0135 \text{ Kg/s/m}^2) \times (44.41 \text{ m}^2) = 0.6 \text{ Kg/s}$$

The evaporation rate will continue to increase as more liquid is released and the puddle area increases. However, at some point, the puddle area may be sufficient to support an evaporation rate higher than the release rate, causing a short-lived peak until an equilibrium between puddle area and release rate is reached.

In this example, the dispersion of the ammonia plume emanating from the vaporizing puddle would then be estimated using a standard Gaussian plume model, since the ammonia vapor is most likely neutrally buoyant. The lower molecular weight of ammonia (17 g/mole) relative to air (28 g/mole) would tend to offset the effect of the cooler plume (initially released at the boiling point) in regard to its density. The downwind distance where the ammonia plume is dispersed to below some critical health-based concentration such as the ATC under conservative meteorological conditions (e.g., F stability class and 2 m/s wind speed) is commonly determined. If the criterion concentration has a relatively long averaging time associated with it (e.g., 1 hour), the equilibrium ammonia emission rate, corresponding to the release rate in this case (10.8 Kg/s), would be appropriate to use. The continuous release Gaussian plume model would be used in this case. If a shorter averaging time concentration is of concern, the peak emission rate corresponding to the maximum puddle area may be more appropriate. Depending on the downwind distances involved, the dispersion of a short-term release corresponding to the peak emission rate may be more appropriately treated as an instantaneous puff release in the Gaussian model.

Chlorine Cylinder Accident

This example deals with the accidental rupture of a 1-ton chlorine cylinder. The chlorine is stored under pressure at ambient temperature to maintain the liquid state, since its boiling point ($-34°C$) is well below ambient temperatures. The rupture of the cylinder is assumed to be severe enough to result in an instantaneous release of the chlorine to the environment. Upon release, a portion of the liquid chlorine instantly flashes to vapor, because it is stored at ambient temperature and its boiling point is well below ambient temperatures. Assuming an ambient temperature of 20°C, the fraction of chlorine flashed to vapor is estimated using equation 10-5 as follows:

$$M_v/M_o = (0.226 \text{ cal/gC})/(68.8 \text{ cal/g}) \times (20°C - (- 34°C)) = 0.18$$

Thus, 18 percent of the 2,000 lb of chlorine released, or 360 lb (163 Kg) forms a puff of chlorine vapor that is subsequently transported and dispersed downwind. The remaining 1,640 lb of liquid chlorine (744 Kg) is assumed to spill unconfined on the ground, forming a liquid pool that vaporizes. The vaporization of the chlorine from the pool results in a ground-level continuous plume that follows behind the instantaneous puff release.

The initial size of the puff release can be estimated from the puff density using equation 10-6, assuming that the puff is at its boiling point temperature after release and that no ambient air is entrained into the puff. Although some air will probably be entrained into the cloud, this assumption is conservative from the point of view of initial puff concentration (pure chlorine). The radius of a spherical puff is then

$$r = [3/4 (163 \text{ Kg})/(3.68 \text{ Kg/m}^3)/3.14]^{1/3} = 2.19 \text{ m}$$

The size of the puddle and its vaporization rate can be estimated in the same manner as that shown in the previous example, since chlorine is also a low-boiling-point liquid.

The dispersion of the instantaneous puff and ground-level plume generated by the vaporizing puddle should be handled using a model with provisions for dense gas dispersion given the high molecular weight of chlorine (70.9 g/mole) relative to air (28 g/mole). Initially, both the puff and plume will also be much colder than ambient air leading to a heavier cloud. The averaging time of the concentration used as a health criterion for the analysis will determine the relative importance of the puff and continuous plume releases. The puff concentration will be critical for short averaging times (several minutes), and the plume concentration will be important for longer averaging times (1 hour or greater). The worst-case meteorological condition for dispersion purposes will be one of low wind speed (< 2 m/sec) and low atmospheric turbulence (F stability class). Although higher wind speeds will cause more rapid vaporization, the worst-case dispersion condition will generally control the analysis. However, it would be prudent to check a few higher wind cases to ensure that conservative results are obtained.

REFERENCES

1. Kaiser, G. D., and Walker, B. C. 1978. Releases of anhydrous ammonia from pressurized containers—The importance of denser-than-air-mixtures. *Atmospheric Environment* 12:2289-2300.
2. Perry, R. H., Green, D. W., and Maloney, J. O. 1984. *Perry's Chemical Engineers Handbook,* 6th ed. New York: McGraw-Hill.
3. Fauske, H. K. 1985. Flashing flows or some practical guidelines for emergency releases. *Plant/Operations Progress* 4:132-134.
4. Leung, J. C. 1986. A generalized correlation for one-component homogeneous equilibrium flashing choked flow. *AIChE Journal* 32:1743-1746.
5. Hanna, S. R., and Drivas, P. J. 1987. *Guidelines for Use of Vapor Cloud Dispersion Models.* New York: Center for Chemical Process Safety of the American Institute of Chemical Engineers.
6. Van Ulden, A. P. 1974. *On the Spreading of Heavy Gas Released Near the Ground.* First International Loss Symposium, pages 211–216. The Hague, Netherlands.
7. New Jersey Toxic Catastrophe Prevention Act, 10 N.J.R. 1356. 1988.
8. Wing, J. 1979. *Toxic Vapor Concentrations in the Control Room Following a Postulated Accidental Release.* Nuclear Regulatory Commission NUREG-0570. Washington, D.C.
9. Shaw, P., and Briscoe, F. 1978. *Evaporation from Spills of Hazardous Liquids on Land and Water.* SRD R 100 UKEA, Culcheth, U.K.
10. Bolz, R. E., and Tuve, G. L. 1973. *Handbook of Tables for Applied Engineering Science.* Cleveland, OH: CRC Press.
11. Pasquill, F. 1974. *Atmosphere Diffusion,* 2nd ed. New York: Halsted Press.
12. Ooms, G. 1972. A New Method for the Calculation of the Plume Path of Gases Emitted by a Stack. *Atmospheric Environment* 6:899-909.
13. McNaughton, D. J., Atwater, M. A., Bodner, P. M., and Worley, G. G. 1986. *Evaluation and Assessment of Models for Emergency Response Planning.* Prepared for CMA by TRC, 800 Connecticut Blvd., East Hartford, CT.
14. Eidsvik, K. J. 1980. A model for heavy gas dispersion in the atmosphere. *Atmospheric Environment* 14:769-777.
15. Woodward, J., Ed. *International Conference on Vapor Cloud Modeling,* Cambridge, MA: The Center for Chemical Process Safety, AIChE. 1987.
16. Colenbrander, G. W. 1980. *A mathematical model for the transient behavior of dense vapor clouds.* Proceedings of the Third International Symposium on Loss Prevention and Safety Promotion in the Process Industries. Base, Switzerland.
17. Colenbrander, G. W., and J. S. Puttock, 1983. Dense gas dispersion behavior: experimental observations and model developments. Proc. 4th International Symposium on Loss Prevention and Safety Promotion in the Process Industries, Harrogate, England.

18. Spicer, T. D., Havens, J. A., Tebean, P. A., and Key, L. E. 1986. DEGADIS—A Heavier-Than-Air Gas Atmospheric Dispersion Model Developed for the U.S. Coast Guard. Paper 86-42.2, *Proceedings of the Air Pollution Control Association Annual Conference,* Minneapolis, MN.
19. Turner, D. B. 1970. *Workbook of Atmospheric Dispersion Estimates.* PHS Publication No. 999-AP-27. Cincinnati, OH: U.S. Dept. of Health, Education, and Welfare, National Air Pollution Control Administration.
20. Accuracy of Dispersion Models. 1977. A Position Paper of the AMS 1977 Committee on Atmosphere Turbulence and Diffusion. *Bulletin of American Meteorological Society* 59 (8), August 1978.
21. Fryer, L. S., and Kaiser, G. D. 1979. *DENZ—A Computer Program for the Calculation of the Dispersion of Dense Toxic or Explosive Gases in the Atmosphere.* SRD R 152 UKAEA, Culcheth, U.K.
22. Technica, Ltd. 1985. *Manual of Industrial Hazard Assessment Techniques.* Prepared for the World Bank. London.
23. Environmental Protection Agency. 1989. *User's Guide for the DEGADIS 2.1 Dense Gas Dispersion Model.* Prepared by Tom Spicer and Jerry Havens, EPA-450/4-89-019.

11

Explosion and Fire Analysis

Frank Elia

The chemical process industry is becoming increasingly interested in the potential effects of chemical plant accidents on plant personnel and the inhabitants of the area surrounding the plant. The effects of fire and explosion hazards are discussed in this chapter. Subsequent plant damage and health effects can be inferred from the fire and explosion analysis results discussed at the end of this chapter.

ANALYSIS METHODOLOGY

Identifying the Accident Scenarios

The first step in performing the explosion and fire analysis is to identify the fire or explosion scenarios to be analyzed. Common screening techniques to define combustible chemical fire or explosion events to be analyzed include failure modes and effects analysis (FMEA), hazard and operability studies (HAZOP), and fault tree analysis (FTA), each of which is discussed in earlier chapters. The screening consists of one or more fire or explosion events resulting from catastrophic component rupture (as of a tank), or a leak at one of several possible locations throughout the plant. In performing fire or explosion analysis, the analyst should evaluate the possibility of secondary fires or explosions resulting from the intense heat released at the primary fire site or the pressure wave generated by the primary explosion site. Each event to be analyzed results in a pool fire, a boiling liquid expanding vapor explosion (BLEVE) or fireball, or an unconfined vapor cloud explosion (UVCE).

Combustible chemicals most frequently evaluated in fire and explosion analysis include but are not limited to the following:

n-Butane	Hydrogen
Butylene	Methane
Ethane	n-Pentane

Ethylene   Propane
n-Heptane   Propylene
n-Hexane   TNT

Release Flow Rates

For postulated piping ruptures in medium- or high-pressure systems (above 2 atmospheres), sonic flow may occur and the chemical release rate will be limited by the flow area at the break location or a smaller flow area away from the break. For failed components such as pumps, valves, and pipe flanges, much lower "leakage" rates occur from flange, stem, or shaft leakage. In low-pressure systems, the break flow is determined by the hydraulic resistance of the system. A catastrophic failure such as a storage tank failure that results in the tank breaking into several pieces is sometimes postulated.

Chemical release rates are determined by a thermal hydraulic analysis, which considers piping resistances, component pressure drop characteristics, pump performance characteristics, and chemical phase change effects. For low-pressure systems with few flow branches and no phase changes following the pipe rupture, a simple hand calculation will do. Most technical data are available in Crane[1]. For medium- and high-pressure systems, phase change may occur within the system. To assess the chemical release rate, a computer analysis may be required. A computer analysis may also be required to account for automatic control logic in the system. The hydraulic analysis must also consider system automatic control logic and operator action that may mitigate or increase the amount of chemical released to the environment. Releases that result from component leakage should be based on historical leakage data for the type of component (e.g., pump, valve) assumed to be leaking. Releases from catastrophic failures are assumed to be instantaneous.

Vapor, Liquid, or Solid

The release rate and phase of the chemical release determines the type of fire or explosion hazard. Extremely rapid release of a liquid stored above its boiling point can result in a BLEVE. A gas leak results in a slow buildup of chemical in the atmosphere. This chemical "cloud" moves away from the plant at a rate that depends on the local weather conditions, including wind speed. Plume dispersion is discussed in detail in Chapter 10. Eventually, sufficient chemical accumulates so that ignition results in an explosion. This is the unconfined vapor cloud explosion (UVCE). Gaseous release can also result from liquids normally at high pressures and temperatures that flash to a vapor-liquid two-phase mixture upon venting to the atmosphere. Liquids with atmospheric boiling points below the temperature of the surrounding environment will vaporize upon leaking to the environment by absorbing energy from the surrounding air. Incomplete vaporization of these liquids results in both a chemical cloud and a liquid pool, thus making it possible for both a pool fire and a UVCE to occur. Release of a liquid with an atmospheric boiling point above the ambient temperature results in a liquid pool that is susceptible to pool fires.

Evaluation of explosion hazards of solids is generally limited to munitions plants and is, therefore, not applicable to the chemical process industry. However, explosion analyses for UVCE relate explosion magnitude in terms of TNT equivalent. Much information has been compiled with regard to TNT explosion predictions and effects on structures by the United States Armed Forces[2].

Generation of Missiles

Catastrophic failures of tanks, rotary equipment, and valves can result in equipment fragments being propelled to other areas of the plant. These airborne fragments may attain sufficient velocity to puncture another tank or cause other plant damage, depending on the size of the fragment and the distance it is propelled. A significant knowledge base for fragment ballistics has been generated by the National Aeronautical and Space Administration[3,4].

Other mechanisms for cascading failures are container failure resulting from excessive flame heat flux or explosion pressure wave.

FIRE AND EXPLOSION CHARACTERISTICS

Boiling Liquid Expanding Vapor Explosion (BLEVE)

A BLEVE can result from a rapid release and vaporization of a volatile chemical following a catastrophic rupture of a tank (e.g., from overpressurization) that contains a chemical with an atmospheric boiling point below the ambient temperature. The intense thermal radiation of a BLEVE is its major danger. Pressure effects beyond the near field are not important. Generally, only a few fragments are generated, but they are large and can travel up to 1,000 meters. The characteristics of a fireball can be estimated with the models presented by Moorhouse and Pritchard[5], and Roberts[6]. These formulas are based on estimates of actual incidents.

The radius of the fireball, r, in meters is given by

$$r = 2.665m^{0.327}$$

where the mass of the chemical released, m, is given in kilograms. The duration of the fireball in seconds, t, can be calculated by

$$t = 1.089m^{0.327}$$

The radiant energy released by the fireball in joules, Q, is

$$Q_R = H_c m \eta$$

where H_c is the heat of combustion in J/Kg and the radiant fraction, η, is given by Roberts[7] as

$$\eta = 0.27P_o^{0.32}$$

where P_o is the initial pressure of the stored chemical in MPa. The radiant heat flux away from the fireball can be estimated by the formula

$$q_R = Q_R/A_r$$

where A_r is the hemisphere area with radius r. Fireball predictions for a typical propane tank failure are shown in Figure 11-1.

Pool Fires

Spillage of liquid chemicals for which the atmospheric boiling point is less than ambient temperature may result in a pool fire. The resulting pool fire radius, r, in meters, can be estimated by the model presented in Shaw and Briscoe[8]:

$$r = (t/\beta)^{0.50}$$

where t is the time in seconds and $\beta = (\pi \rho_L/8 \ gm)^{0.50}$ where

$$\rho_L = \text{chemical density, Kg/m}^3$$
$$g = \text{gravitational constant, m/sec}^2$$
$$m = \text{mass of spilled chemical, Kg}$$

For continuous spills the pool radius is given by

$$r = (t/\beta)^{0.75}$$

where $\beta = (9\pi \rho_L/32 \ \dot{g}\dot{m})^{0.333}$ with the rate of chemical release, $\dot{m}$, in Kg/sec.

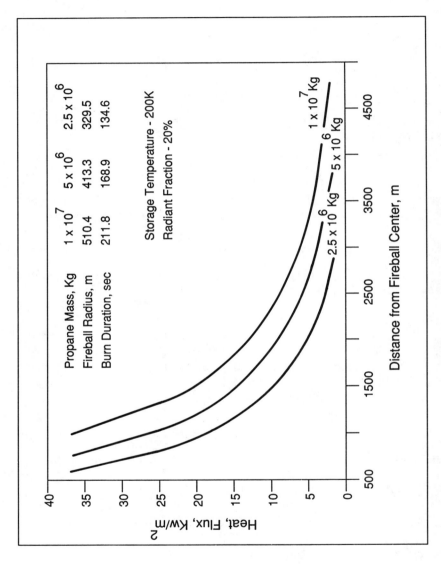

Figure 11-1 Propane tank failure.

At any time, the pool spreading is caused by gravity and is circular with uniform height. The pool continues to spread until it reaches an enclosing bund wall or other barrier.

Evaporative losses from the pool can be calculated by Sutton's equation[9] if the ambient temperature is below the atmospheric boiling point:

$$M_\iota = a(P_sM/RT_A)u^{(2 - n)/(2 + n)}r^{(4 + n)/(2 + n)}$$

where a is 4.78×10^{-3} to represent neutral atmospheric stability and n is set equal to 0.25, the midpoint of the observed range and

P_s = saturation vapor pressure, n/m^2
M = molecular weight
R = universal gas constant
T_A = ambient temperature, °K
u = wind speed at 10 m, m/sec

If the dominant source of heat is conduction from the ground, then the model proposed by several workers[10,11,12] is appropriate:

$$M_\iota = \pi r^2 k(T_A - T_B)/(H_{vap}t^{1/2})$$

where

k = average soil constant (6.68×10^{-5})
T_B = atmospheric boiling point, °K
H_{vap} = heat of vaporization, J/Kg

Ground conduction is dominant for about 1 minute; the exact time is dependent on wind speed.

If the pool burns, the evaporation rate is replaced by the burning rate. The burning rate, $\dot{m}_c$, for chemicals with boiling points above the ambient temperature is given by the model proposed in reference 13:

$$\dot{m}_c = \frac{0.001H_c\pi r^2}{C_p(T_B - T_A) + H_{vap}}$$

where

C_p = Heat capacity of the chemical, J/Kg − °K

For chemicals with boiling points below the ambient temperature,

$$\dot{m}_c = \frac{0.001 H_c \pi r^2}{H_{vap}}$$

The flame height, H, in meters, is calculated by

$$H = \frac{84 r [\dot{m}_c (2gr)^{0.50}]^{0.61}}{\rho_A}$$

with ρ_A equal to the density of air.

The radiant heat released from the pool fire is calculated by

$$Q_R = \frac{(\pi r^2 + 2\pi r H) \dot{m}_c \eta H_c}{72 \dot{m}_c^{0.61} + 1}$$

For pool fires, the radiant fraction η has been shown in tests to have values up to 0.35. Pool fire example results for n-heptane are presented in Figure 11-2.

Unconfined Vapor Cloud Explosion (UVCE)

Continuous release of combustible gas into the atmosphere could result in the formation of a cloud. The cloud passes over the plant grounds and the surrounding area carried by the winds. High wind speeds promote dispersal of the gas, limiting gas concentrations to below the lower explosive limit (LEL), thereby eliminating the possibility of combustion or explosion. This is the lower limit of chemical concentration for which combustion is possible. The upper or higher explosive limit (HEL) is the highest concentration for which combustion is possible. Values of explosive limits can be found in the *Handbook of Chemistry and Physics*[14], for various substances. A low enough wind speed could result in a chemical cloud formation and possible movement over the surrounding areas outside the plant. Vapor clouds are represented as shown in Figure 11-3. In UVCE analysis, the ellipsoid surfaces are determined for which the airborne chemical concentrations are equal to the HEL and the LEL, respectively. The concentration is integrated over the volume contained within these two ellipsoid surfaces to obtain the detonable mass.

First the horizontal and vertical dispersion coefficient, σ_y and σ_z, are calculated as a function of x, the distance downwind from the chemical release point. The centerline maximum volume concentration is then calculated based on the work presented in Burgess and Zabetakis[15]:

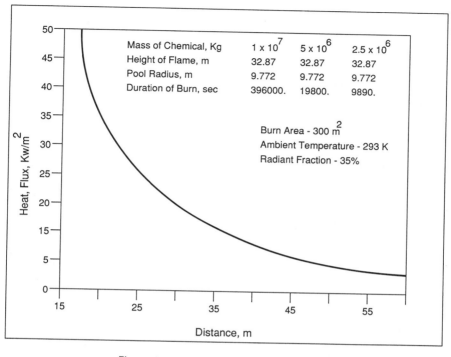

Mass of Chemical, Kg	1×10^{7}	5×10^{6}	2.5×10^{6}
Height of Flame, m	32.87	32.87	32.87
Pool Radius, m	9.772	9.772	9.772
Duration of Burn, sec	396000.	19800.	9890.

Burn Area - 300 m^2

Ambient Temperature - 293 K

Radiant Fraction - 35%

Figure 11-2 Pool fire study for n-Heptane.

$$\chi_{CL} = Min \left(\frac{Q}{3.53 \, \sigma_y \, \sigma_z u}, \, 1.0 \right)$$

where Q = the release rate of flammable gas in ft³/sec.

The major and minor axes which correspond to the HEL ellipsoid surface are calculated by

$$Y_{HEL} = \begin{cases} 0, \, \chi_{CL} \leq HEL \\ \sigma_y \, [2 \, \ln(\chi_{CL}/HEL)]^{1/2}, \, \chi_{CL} > HEL \end{cases}$$

and

$$Z_{HEL} = \begin{cases} 0, \, \chi_{CL} \leq HEL \\ \sigma_z \, [2 \, \ln(\chi_{CL}/HEL)]^{1/2}, \, \chi_{CL} > HEL \end{cases}$$

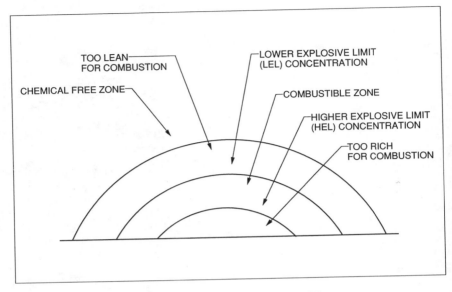

Figure 11-3   Cloud model for UVCE.

Likewise, the major and minor axes are calculated for the LEL ellipsoid as follows:

$$Y_{LEL} = \begin{cases} 0, \chi_{CL} \leq LEL \\ \sigma_y \, [2 \, \ln(\chi_{CL}/LEL)]^{1/2}, \chi_{CL} > LEL \end{cases}$$

and

$$Z_{LEL} = \begin{cases} 0, \chi_{CL} \leq LEL \\ \sigma_z \, [2 \, \ln(\chi_{CL}/LEL)]^{1/2}, \chi_{CL} > LEL \end{cases}$$

The average concentration in the flammable zone of the cloud is

$$\bar{\chi} = \frac{2}{A} \int_0^{\pi/2} \int_{R_{HEL}}^{R_{LEL}} \chi r \, dr \, d\Theta$$

where

$$A = \frac{\pi}{2} \, (Y_{LEL} \, Z_{LEL} - Y_{HEL} \, Z_{HEL}), \, m^2$$

$$\chi = \chi_{CL} \exp \left[\frac{-r^2}{2} \left(\frac{\cos^2\Theta}{y^2} + \frac{\sin^2\Theta}{z^2} \right) \right]$$

$$R_{HEL} = \left(\frac{\cos^2\Theta}{Y_{HEL}^2} + \frac{\sin^2\Theta}{Z_{HEL}^2} \right)^{-1/2}$$

$$R_{LEL} = \left(\frac{\cos^2\Theta}{Y_{LEL}^2} + \frac{\sin^2\Theta}{Z_{LEL}^2} \right)^{-1/2}$$

The expression for χ has been transformed from Y-Z coordinates to r-Θ coordinates by substituting

$$Y = r \cos\Theta$$
$$Z = r \sin\Theta$$

The average concentration is found by integrating the double integral above to give

$$\bar{\chi} = \frac{2\sigma_y\sigma_z}{Y_{LEL}Z_{LEL} - Y_{HEL}Z_{HEL}} (\exp - Y_{HEL}^2/2 \ \sigma_y^2 - \exp - Y_{LEL}^2/2 \ \sigma_y^2)$$

The incremental volume of the flammable cloud is calculated from

$$\Delta V = 35.3 \ A \ \Delta x, \ ft^3$$

The incremental equivalent weight of TNT is calculated by

$$W = \bar{\chi}\Delta V \ \rho_g \ H_c \ / \ 3.968 \times 10^6, \ tons$$

Volumes and weight increments are added for downwind distance increments, x, until the value of the centerline concentration falls below the LEL.

The radial distance from the cloud center to the location of incident pressure is calculated from

$$R = \lambda \ W^{1/3}, \ m$$

for which W is in tons of TNT. Values of incident pressure versus λ are specified in the following:

INCIDENT PRESSURE (psi)	λ
10.0	30.48

5.0	45.72
3.5	57.91
3.0	68.58
2.5	83.82
2.0	103.60
1.5	140.20
1.0	213.40
0.8	274.30
0.5	518.20

For example, if one wished to determine the distance from the cloud center to the incident pressure point for 3.0 psi or less, the above equation would be used substituting 68.58 for λ.

A pressure map can then be superimposed on the plant area to estimate plant damage and personal injuries. These pressure isobars are drawn concentrically about the blast center. Pressure results are summarized for a propane release in Figure 11-4.

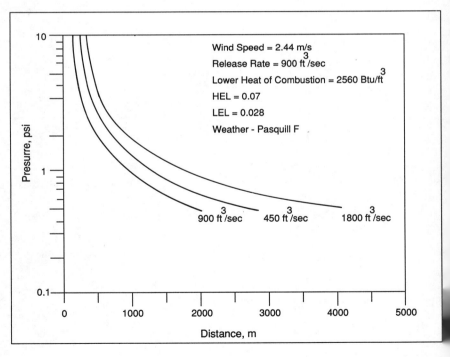

Figure 11-4   UVCE overpressure results for propane.

specifically, the flash point, the lower explosive limit (LEL), and the upper explosive limit (UEL). The flash point of a substance is the lowest temperature at which a liquid will produce enough vapor to create an ignitable mixture in air (at standard pressure). The LEL is the minimum concentration of vapor in air necessary to support combustion, while the UEL is the highest vapor concentration in air that can support combustion. In other words, the explosive range is between the UEL and the LEL. The UEL is also sometimes referred to as the highest explosive limit, or HEL.

Assessing the Magnitude of a Potential Release

The next step in assessing health effects from chemical releases is to assess the magnitude of the potential release. *Magnitude* is used here as a broad term that includes the amount of release, usually given in terms of lb/min, the surrounding population, atmospheric conditions that may affect chemical dispersion, and size of the affected area.

To determine the rate of release, the protocol in the EPA/FEMA/DOT's guide, *Technical Guidance for Hazards Analysis*[5], will be used. To estimate the potential release as outlined in the guide, the physical state of the release must be known. For gases, the rate of release is determined by dividing the quantity of the substance that could be released from a vessel within 10 minutes by 10, which gives a result in lb/min. Particulate and liquid release rates are calculated using the same method. It is important to remember that this is a release rate calculation, not a concentration calculation. The concentration of a release at a given location is dependent on atmospheric conditions and terrain. This topic is covered in detail in Chapter 10, ''Chemical Plume Dispersion Analysis.''

The calculations for rate of release for vapors are not as straightforward as those for gases, liquids, and particulates. This is because the liquid from which the vapor is derived may not volatilize immediately. Vapor release rates, as well as vapor concentrations, depend on atmospheric conditions and are also discussed in Chapter 10.

The demographics of the surrounding population is another consideration that contributes to the magnitude of the release. Children, the sick, and the elderly are more apt to experience adverse health effects from chemical exposure than healthy young adults, so it is important to note the proximity of schools, hospitals, nursing homes, and recreation areas to the potential release site.

Atmospheric conditions also have an effect on the magnitude of a release. The wind speed and direction as well as other parameters have a great impact on how a release travels. This topic is discussed in Chapter 10.

All these factors contribute to the magnitude of the release and play an important part in assessing health effects.

Determining Toxicity

The third step required to assess health effects from chemical releases is to determine the toxicity and explosivity of the chemical. Only toxicity will be addressed in this section as Chapter 11, "Explosion and Fire Analysis," discusses explosivity.

Toxicity can be defined as the ability of a substance to adversely affect the health of an organism: the higher the toxicity of a substance, the more hazardous it is to the health of human beings or other animals. Toxicity is also affected by the route of exposure (e.g., inhalation, ingestion, and skin contact). Some chemicals are toxic to ingest, but not to touch.

Toxicity can be measured in many ways. Some common measures are defined and discussed in the following paragraphs.

Immediately dangerous to life and health (IDLH) is a term that was defined by the Standards Completion Program (SCP), a committee with representatives from both the Occupational Safety and Health Administration (OSHA), and the National Institute of Occupational Safety and Health (NIOSH), for the purpose of respirator selection. IDLH represents the maximum concentration to which a worker could be exposed for 30 minutes and experience no permanent adverse health effects.

Median lethal dose (LD50) is the amount of chemical that kills 50 percent of the exposed test population. The routes of exposure for lethal dose tests are ingestion or skin absorption, and the amount is usually expressed in milligrams of substance per kilograms of test subject body weight (mg/kg).

Median lethal concentration (LC50) is the concentration of chemical that kills 50 percent of the exposed test population through inhalation.

Lethal dose low (LDLO) is a term related to the toxicity of a substance, and refers to the lowest dose required to kill some of the test population through ingestion or skin absorption, while *lethal concentration low* (LCLO) is the lowest concentration required to kill some of the test population through inhalation.

Acute toxicity concentration (ATC) refers to the lowest confirmed concentration level that caused death or permanent disability to the test subject.

Permissible exposure limit (PEL) and time-weighted average-threshold limit value (TWA-TLV) are also sometimes used to assist in determining the toxicity of chemicals. Both terms give a level of contaminant to which an average healthy worker can be exposed for 8 hours a day, five days a week, without experiencing adverse health effects. These values, however, should not be used as the sole determination of the toxicity of a substance. If a release were to occur, exposure of children, the sick, and the elderly would also occur, and these sections of the population would probably be more sensitive to exposures than the healthy working individuals for whose protection these standards were designed.

Other yardsticks to reference toxicity include the Substance Hazard Index (SHI). This value is used by the state of California (CA) in the document Guidance for the Preparation of a Risk Management and Prevention Program and by OSHA in their notice of proposed rule-making concerning process safety management of highly hazardous chemicals. The SHI is defined as

$$(CA) \ SHI = \frac{\text{equilibrium vapor concentration (ppm @ 25°C)}}{\text{acute toxicity concentration}}$$

or

$$(OSHA) \ SHI = \frac{\text{equilibrium vapor concentration}}{\text{emergency response planning guidelines (III)}}$$

Emergency response planning guidelines provide the maximum airborne concentration below which nearly all individuals could be exposed, for up to 1 hour, without experiencing or developing life-threatening health effects and are published by the American Industrial Hygiene Association (AIHA).

In both regulations, the equilibrium vapor concentration is defined as

$$\frac{\text{vapor pressure @ 20°C} \times 10^6}{\text{total atmospheric pressure}}$$

Using these terms can give a quantitative estimate of toxicity.

Estimating Exposure

The fourth step in the assessment process is to estimate exposure to the surrounding population. Exposures are classified as either acute or chronic. Chronic exposures are exposures that occur over a long period of time in small concentrations. Acute exposures are more of a concern in the case of chemical releases and are defined as exposures that occur from a sudden event and are of relatively high concentrations.

The exposure of any area is determined by the concentration of chemical in the air at that location and the mode of exposure. The level of concern, or LOC, of the chemical itself is also a factor in determining exposure. The level of concern is "the concentration of an extremely hazardous substance in air above which there may be serious irreversible health effects or death as a result of a single exposure for a relatively short period of time"[5] (USEPA, FEMA, USDOT, 1987). Currently there is no rule of thumb for calculating the LOC. It can, therefore, be left to the individual performing the risk assessment to choose the basis for the calculations. For example, the level of concern chosen could be the value listed in the Emergency Response Planning Guidelines or it could be one-hundredth of the LD50 value for a specific chemical. Again, it must be emphasized, there is no recognized method for calculating a LOC.

Exposure can also occur through ingestion of contaminated water or food. Again, the concentration of the contaminant must be estimated.

Finally, exposures can be estimated from sampling data from previous releases, including air sampling, soil sampling, water sampling, and swipe sampling.

Characterizing Health Risks

The fifth and final step in the assessment process is to use the information obtained from the previous steps to characterize the health risk. In other words, this step is the actual estimate of health effects experienced by the surrounding population. One general method of estimation is the comparison of the concentration level of the area of interest to the LOC. If the concentration level of the area of interest has the potential to be higher than the LOC for that contaminant, the EPA recommends that the area should be targeted for evacuation, because any concentration above the LOC could cause irreversible damage to the health of the surrounding population.[1]

A more specific method is to examine the dose received by any individual in the exposed area. The calculation for the dose received depends on the route of exposure. For example, for inhalation, the dose received would be

$$IEX = \frac{(d)\ (I)\ (C)\ (F)}{(BW)\ (25600\ days/lifetime)}$$

where

IEX = estimated inhalation exposure (mg/kg day)
d = duration of event (hours/event)
I = average inhalation rate of exposed individuals (m^3/ hour—can be conservatively estimated to be 1 m^3/ hour)

C = contaminant air concentration (mg/m^3)
BW = body weight (kg—can be conservatively estimated to be 70 kg)
F = frequency of exposure $\left(\dfrac{event}{lifetime}\right)$

The response of the individual to the dosage received can be found by comparing the result obtained to dose and response data of the chemical of interest.

The preceding five steps give a basic method to follow to assess potential effects from chemical releases. Anyone using the method must remember that the objective of the exercise is to estimate the health effects from a chemical release if it were to occur. Results obtained from a risk assessment are for the purposes of prevention planning only. If a release were to occur, data from the actual release would have to be used to determine health effects and evacuation zones.

THE PRIORITIZATION PROCESS

The assessment process can be a useful tool in assessing health effects from chemical releases. Unfortunately, it can be a labor-intensive process if several different chemicals are present at a facility. The chemicals present need to be prioritized according to the potential for release[7]. Certain chemicals exhibit properties that give them a greater potential for release. Based on those properties, a process of elimination can be used to narrow down the number of different chemicals that need to be evaluated. Some of the properties to be considered are

- Listing as a hazardous substance by a federal or state regulatory agency
- Physical state of substance
- Quantities handled or stored
- Toxicity
- Flammability
- Reactivity
- Decomposition products
- Combustion products
- Potential for inadvertent combination with materials stored nearby
- Storage considerations (pressure and temperature).

Although definitive limits do not exist for each parameter to determine whether a substance has a higher or lower potential for release, the following can be used as a guideline. A chemical under consideration should be included in the risk assessment if it meets any of the following conditions:

- It is a gas or vapor or it decomposes to emit a gas or vapor.
- The amount handled or stored meets or exceeds a reportable quantity.
- It emits hazardous decomposition products.
- It is reactive.
- It is toxic.
- It is flammable or combustible.
- The storage temperature exceeds the boiling point of the chemical.

Toxicity and flammability can be further broken down. The cutoff values for toxicity can be based entirely on lethal dose or lethal concentration values, or it can be based on the Material Hazard Index (MHI) or Substance Hazard Index (SHI). The formula for the SHI was previously given in the "Assessment Process" section. The MHI is

$$MHI = \frac{equilibrium\ vapor\ concentration\ @\ 25°C}{level\ of\ concern}$$

For reference, a value of 1388 for the MHI was chosen by the state of New Jersey for its Toxic Catastrophe Prevention Act Program[8]. This represents the MHI for a 36 percent solution of hydrochloric acid. An MHI above that value would represent a chemical to be included in the risk evaluation process.

The cutoff for flammability is determined by the probability that the chemical could ignite in ambient conditions. Any chemical with a flash point less than 93.3°C (200°F) should be included. If flash point data cannot be found, the rating assigned to the chemical by the National Fire Protection Association (NFPA) can be used. In this case, any liquid chemical with an NFPA rating of 2 or higher or any solid with a rating of 1 or higher should be included.

Prioritizing the chemicals by potential for release will narrow down the number of chemicals that need to be evaluated. Another effective method of limiting the number of chemicals needing evaluation is through chemical selection, which is a preventive measure[9].

THE PREVENTION PROCESS

The prevention process works on the assumption that if less hazardous chemicals are used to begin with, there will be fewer chemicals to be evaluated for potential adverse health effects in the event of a release.

Based on this assumption, it is logical to state that, whenever feasible, chemical hazard is one of the parameters to consider when selecting chemicals to be used in a chemical process. The same guidelines that are used in the prioritization process can be used in the selection of chemicals. There are more advantages to selecting less hazardous chemicals than in limiting their potential for release; for example, a chemical that is less hazardous as virgin material may be less costly to treat once it is accidentally released.

When chemical substitution is not feasible, another option is to keep inventory to a minimum, because a chemical stored in a less-than-reportable quantity may be eliminated from the assessment process.

SOURCE OF INFORMATION

The preceding processes demonstrate a basic methodology of performing health risk assessments. Performing these processes requires gathering a significant amount of data. Several references exist that provide the needed information, both in literature and through computer data bases. The objective of this section is to evaluate and list some of these sources.

One very useful source of information is the Material Safety Data Sheet (MSDS). There should be an MSDS for most chemicals encountered, as it is a requirement of OSHA 29 CFR 1910.1200, otherwise known as the Hazard Communication Act or Right-to-Know Law, that chemical manufacturers and importers "obtain or develop a MSDS for each hazardous chemical they produce or import." [10] An MSDS is also a requirement of the Superfund Amendments and Reauthorization Act of 1986 (SARA Title III).

A comprehensive MSDS typically contains much of the information mentioned in the previous sections, including

- Chemical manufacturer's name, address, and emergency address
- Components of the mixture
- Permission exposure limit (PEL) or threshold limit value (TLV)
- Physical and chemical data
- Flammability limits
- Hazardous decomposition products
- Reactivity data
- Handling procedures
- Spill response procedures
- Health effects
- First-aid data
- Control measures.

Material Safety Data Sheets are available from the manufacturer of the chemical. In addition, there are a number of other sources from which an MSDS can be obtained. One such source is

Genium Publishing Corporation
1145 Catalyn Street
Schenectady, NY 12303-1836
(518) 377-8855.

In addition to MSDS, there are several other good sources of information. A few of them are the following:

SOURCE	DESCRIPTION
Dangerous Properties of Industrial Materials, 6th ed., I. N. Sax, 1984, Van Nostrand Reinhold	Listings are by chemicals. Gives physical and chemical data, synonyms, CAS number, toxicity data, potential for disaster hazard.
Dangerous Properties of Industrial Materials, 4th ed., I. N. Sax, 1975, Van Nostrand Reinhold	Gives same information as 6th edition except for toxicity. Does not have as many listings, but has useful chapters on toxicology, fire protection, and so forth.
NIOSH Pocket Guide of Chemical Hazards, U.S. Dept. of Health and Human Services, Publication 85-114, Available by calling NIOSH (513)533-8257	Excellent source. Gives listings by chemical and information on synonyms, exposure limits, IDLH level, physical description, chemical and physical properties, incompatibilities.
Handbook of Toxic and Hazardous Chemicals and Carcinogens, 2d ed., M. Sittig, 1985, Noyes Publications	Listing by chemical. Gives physical and chemical data, common uses of chemical, effects and routes of exposure, synonyms. Has useful index of carcinogens.

DATA BASES

Acute Hazards Data, Available through the Office of Economics and Technology, U.S. EPA.	Gives acute toxicity of various substances.

The Integrated Risk Information System (IRIS) Available through U.S. EPA
Public Health Risk Evaluation Data Base. Available through the U.S. EPA

Chemical Exposure
Available through
Oak Ridge National Laboratory
Environmental Mutagen Information Center,
Building 9224, P.O. Box Y,
Oak Ridge, TN 37831
(615) 574-7871

Hazardline
Available through
Occupational Health Services, Inc.
400 Plaza Drive
P.O. Box 1505
Secaucus, NJ 07094
(800) 223-8978

Listing by chemical. Gives dose-response information, toxicity levels, carcinogenicity bioassays. Personal computer software package. Gives information on chemical, physical, toxicological data and health based standards. Gives effects of toxic chemicals on animals and humans.

Contains regulatory and health information on a large number of chemicals. Also contains physical and chemical data, exposure limits, CERCLA hazard ratings, route of entry of exposure. Very comprehensive.

CONCLUSION

The overall objective of the assessment process is to identify whether an impact on public health exists in the exposed population and, if so, who is affected. A method by which to determine this is presented in this chapter. The information gathered from performing an evaluation can often be essential in making action decisions, such as changing practices in chemical selection, hazardous materials handling, and plant operations. The risk estimate may also be used for decision-making in the management of mitigation of the chemical release. If the risk estimate indicates that an adverse health effect may occur in the exposed population, it can aid in prioritizing the allocation of resources for health-care services. Finally, the results obtained from the risk assessment can be used in the planning stages as an indicator of the potential need for evacuation or sheltering facilities. This topic is covered in Chapter 16, "Emergency Preparedness."

REFERENCES

1. U.S. Environmental Protection Agency. April 1988. *Superfund Exposure Assessment Manual,* 122.
2. U.S. Environmental Protection Agency. November 1986. *Superfund Risk Assessment Information Directory,* 3-1–3-49.
3. U.S. Department of Health and Human Services. February 1987. *NIOSH Pocket Guide to Chemical Hazards,* p. 14.
4. Sax, I. N. 1975. *Dangerous Properties of Industrial Materials,* 4th ed., pp. 246–273. New York: Van Nostrand Reinhold.
5. U.S. Environmental Protection Agency/Federal Emergency Management Agency/U.S. Department of Transportation. December 1987. *Technical Guidance for Hazards Analysis: Emergency Planning for Extremely Hazardous Substances.* pp. A-6, D1-D6, 3-3.
6. California Office of Emergency Services. September 1989. *Guidance for the preparation of a risk management and prevention program:* pp. D-2, D-4, C2-C3.
7. Department of Labor—OSHA. February 1990. *Process Safety Management of Highly Hazardous Chemicals,* (Notice of Proposed Rulemaking), 91–92.
8. State of New Jersey Department of Environmental Protection, Proposed Rule N.J.A.C. Toxic Catastrophe Prevention Act Program: 7:31, p. 3.
9. *Industrial Hygiene Report.* June 1987. New York: American Insurance Services Group, Inc.
10. Department of Labor—OSHA. August 24, 1987. Hazard Communication Act—Final Rule, Federal Register. 163(52):31869-31881.

13

Quantified Risk Assessment

Robert L. O'Mara, Harris R. Greenberg, and
Robert T. Hessian, Jr.

When the results of failure modes and effects analyses (FMEAs), fault trees, hazard and operability studies (HAZOP) criticality analysis, human reliability event trees, and other techniques are quantified with data, a whole new dimension and means of comparison are added to risk assessment. Quantified results or models enable the analyst to compare results in relative terms and to sort or rank them in order of their effect on the plant or the public. The top ranked events or event sequences then become the critical or controlling events or event sequences. It is upon the latter that efforts at reduction or mitigation have the greatest effect.

This chapter provides an overview of the concepts and steps required in the collection, management, and use of quantitative data in risk assessment. For more specific information as to the types and level of data needed, see Chapter 7 for FMEA, Chapter 9 for fault trees, Chapter 8 for HAZOP, and Chapter 14 for human reliability analysis.

BRIEF REVIEW OF DATA APPLICATIONS

Background

Probability is an estimation of the frequency of occurrence for a specific event or failure. The frequency with which this may occur depends on factors such as the number of contributors in a given population (e.g., 30 valves of design X in a plant that utilizes 4,000 valves of that type); the frequency of functional demand upon the component (e.g., a standby pump may be placed in service only five times per year, or a batch feed pump may be operated twice per shift and six days per week); the length of time that the component is required to function once it has been successfully started (e.g., once started, a cooling water system is required to operate for 21 hours continuously during a batch operation); the maintenance program elements that remove the component from service (e.g., disassembling a component for preventive maintenance not only introduces a period of unavailability, but may also introduce latent maintenance errors).

Testing, shared duty, and adverse environmental conditions can also contribute to the frequency of a specific event or failure. Testing frequently involves deliberately imposed high stresses or duty on equipment. Shared duty may mean that transitional stresses are incurred more often than would be required if installed spare equipment were used only when the primary equipment was being repaired. Equipment exposed to extreme temperatures or to a corrosive or dirty environment for prolonged periods may have a shorter life span and a higher frequency of failure.

Thus, failure rate data provide an estimate of the probability of a failure occurring in a specific component or functional group of components over a specified period, either of time or of operating exposure, such as number of cycles. A basis for the collection, assessment, and synthesis of failure data must be established to make data selection and usage consistent with the application.

An example of one such application shows the extent of data needed. The following is a summary description of a batch process that employs four reactors. Each reactor in the process has six chemical feed systems, a jacket heating steam system, a jacket cooling water system, a vent to a common scrubber system, an agitator, two programmable logic controllers, a primary and alternate off-site power source, as well as an on-site emergency power system, and a nitrogen purge system.

This batch reactor system incorporates several other systems, and a complete listing of individual components would be quite detailed and lengthy. Even the software interlocks and error handling protocol could be complex, depending on the nature of the process.

Data Collection and Management

Two types of data are generally available to the analyst: (1) plant specific data and (2) generic data from other sources. Plant specific data that reflect the actual equipment failure frequencies and maintenance practices used are regarded as the most applicable, but for a process that is still essentially on the drawing boards, or one in which no initial data are available, other sources and generic data may be used for a first cut analysis.

Several steps must be taken to accomplish the task of data collection and management. First, the end objective should be identified and clearly described. A list should be developed of all the components to be included in the models to be quantified. This is the list of data needs for the data base.

Second, the required level of detail of each model needs to be determined. Models are usually developed to the level of detail for which data are available, but assemblies of components, such as refrigeration equipment, may make this unnecessary, if only the broad function needs to be accounted for in the model. As a data management tool, the use of assembly levels, when properly carried out, can greatly simplify the task of quantified risk assessment.

Third, the operating exposure period must be established to evaluate the probability of failure of each equipment item or subsystem. The operating exposure may be in terms of operating hours, calendar time, or operating cycles.

Fourth, because operating conditions and degree of restoration during repair can heavily influence the rate of failure, the operating age of an equipment item, the period between failures, the operating mode, maintenance performed, and environmental conditions should also be collected. For example, valve positioners on valves subject to cavitation may fail twice as often as those without this heavy vibrational environment.

Similar information should be made available for all functionally associated pieces of equipment: the quality, type, temperature, velocity, and other characteristics of the process fluid should be noted for their effect on materials of construction and for process design.

Fifth, when preparing to calculate the probability of system failure or unavailability, the failure modes of the various contributing components must be understood. The reason for this level of detail is that some component failure modes can cause the whole system to fail, whereas others only affect or degrade the function of the component itself. These failure modes may be dependent on the operating mode or status of the equipment. For example, valve stem leakage or stem binding may result from the degree to which the gland is tightened. A small amount of leakage will probably not fail the system function, but stem binding may prevent the valve from moving to its desired position.

Operating status may dictate which failure modes apply. For example, systems often contain two 100 percent capacity pumps in which one pump is assigned to operate and the other is in standby. The system analysis should reflect the fact that one pump is already running while the other is not. The probability that the running pump will fail to continue running will be different than for the pump in standby failing to start. Contributors to the probability of failure of the running pump may include excessive leakage such as from seal failure, or a physical fault such as a motor short or a coupling failure, while the contributors to failure to start and run of the standby pump may include an undervoltage relay trip, failure of a starter contact, or that the pump develops insufficient discharge head in time to prevent a protection system shutdown.

Typical Sources of Generic Data

Several sources of generic data are available to the analyst. This section includes a sample list of data sources that is by no means complete or exhaustive. Each set of data was compiled for a specific purpose that may or may not match what the analyst is attempting to describe with an analytical model. For example, IEEE Std 500, 1984[1], originally conceived for use in nuclear plant reliability models, contains a very broad range of data specifications, some of which may be perfectly applicable to process applications and models. Although the data in this case do not specifically include the effects of a chemical environment, environmental factors may be judiciously applied to them.

A pitfall that the analyst should avoid is incompatibility of data format. Failure rate data are usually expressed in terms of failures per million hours of either operating time or calendar time. Demand rate data or failures per demand may be essential for a standby component that operates only for a brief period or in cases in which most of the transients and stressors occur during startup.

The numerical result is also very sensitive to the definition used to describe the data. IEEE Std 763 has attempted to redefine the process of starting a gas turbine as a success or failure in a period of 20 minutes, as opposed to instantaneous success or failure on the first attempt. The result of this redefinition is 98.5 percent success for the 20-minute window, versus 95 percent success for the instantaneous case. System design and the number of parallel units already in operation will dictate which of these definitions is more practical.

Typical sources for component failure data include the following:

1. Reactor Safety Study, WASH-1400. 1975. Although compiled principally for the nuclear power field, much of the data was originally derived from chemical process information.
2. IEEE Standard 500, 1984 *Reliability Handbook*. 1983. New York: Institute of Electrical and Electronics Engineers.
3. Component Failure and Repair Data for Coal-Fired Power Units, EPRI-AP-2071, Electric Power Research Institute. 1981. This has some flue-gas desulfurization equipment included and addresses common failure modes.
4. Imperial Chemical Industries, LTD. Reliability Data, Report No. MD19128/1. 1980. (Proprietary to ICI, but purchasable).
5. Lees, F. P. *Loss Prevention in the Process Industries*, Vol. 2. 1980. London: Butterworths.
6. Green, A. E., and Bourne, J. R. *Reliability Technology*. 1972. New York: John Wiley.

7. Nonelectronic Parts Reliability Data. 1985. IIT Research Institute, Reliability Analysis Center, Rome Air Development Center, Griffiss Air Force Base.
8. Henley, E. J., Kumamoto, H. *Reliability Engineering and Risk Assessment*. 1981. Englewood Cliffs, NJ: Prentice-Hall.

Data Manipulation

Once a source or a few applicable sources of data have been obtained, the analyst's job is not complete. Data sources all have limitations and differences with the design or process under consideration, and some attempt should be made to reconcile the differences or to manipulate the data to more nearly fit the application.

In the previous example with IEEE 500 data, there are environmental factors, such as for vibration and temperature, that can be applied using the values suggested in the handbook.

Where data are severely limited by being derived from only a few data points, pooling of data from different sources may be attempted. A typical case is when plant specific data are dependent on only two or three failures, and by updating a prior distribution of generic data using the Bayesian technique[2] or pooling with other data, the plant data may be made usable. Pooling implies that the distribution of the data from one source is similar to that from another, so that the pooled data distribution is essentially not different. The pooled distribution may be quite different, however.

UNCERTAINTY AND SENSITIVITY ANALYSIS

Background

Uncertainty is a statement of belief in the representativeness of mean-centered values in the results of a quantified risk assessment. Sensitivity analysis attempts to show what would happen to these values if a parameter or series of assumptions used to generate the results were changed. In either case, the techniques involved are used to better describe the state of knowledge of the assessment and the degree of confidence in the results.

Uncertainty analysis is performed to estimate the effects of randomness in the study parameters or input data, and the effects of a lack of precision or completeness in the models used or in the model predictions. Uncertainty in final results can be characterized as additional risk to public health and safety and in frequency of events that pose such a hazard.

Definition of Uncertainty

Conceptually, there are two different categories of uncertainty:

1. A random variability in some parameter or measurable quantity such as input data on failure rates.
2. An imprecision in the analyst's knowledge about the completeness and applicability of models, the precision of their parameters, and the model's ability to correctly predict event outcomes.

Consider an example using light bulbs to illustrate the difference between these two categories:

1. The model has to account for the randomness with which the bulbs burn out in similar applications.
2. The model and its applied data are inadequate to fully account for the dynamics of variables such as operating time, time-temperature history, residual gas content effects, cumulative effects of shock and vibration, and electrical power supply quality, such as voltage regulation.

The precision in the second concept might be improved by collecting more data, along with predictability under controlled conditions. This may reduce the numerical uncertainty or increase the confidence level for the model prediction, but the essential randomness of the first type will still exist. Although inert gas-filled light bulbs generally last longer, the distribution of their filament lives in real applications is still quite broad.

From the preceding example, it is apparent that it is not easy to separate inherent randomness from lack of knowledge. Often, both concepts are combined in the treatment of uncertainty and are estimated by such random sampling techniques as Monte Carlo simulation to produce parameter value distributions. Current usage of "uncertainty analysis [is] understood to mean the analysis of how both random variability in parameter values and uncertainties [from modeling] propagate through [the risk assessment] to give a single uncertainty/variability measure for the results . . ."[3]

The following are practical examples of the types of uncertainty that must be considered[3]:

CATEGORY	EXAMPLES
Parameter	Data may be incomplete or biased. Are all failures listed, and are the number of starts or trials known?

	Do the available data apply in this case? Can generic data be applied here? Is the method of data analysis valid? Do failure modes other than the one in question dominate the failure rate?
Modeling	Is a binary decision model used to represent a multivariate set of cases or a continuous process? What uncertainty is introduced by mathematical conveniences such as log normal distribution? Is the model being used outside the range to which it could be expected to apply?
Completeness	Does the model exclude a whole range of events that are as probable as the range of events that are addressed? Have human errors and common-cause failures been considered? Are all important physical and chemical processes addressed? Will the depth of analysis support the assessment conclusions?

Uses of Sensitivity and Uncertainty Analysis

Uncertainty analysis can be performed either qualitatively or quantitatively, depending on the level of uncertainty analysis needed. In this context, it is suggested that a qualitative uncertainty analysis be performed as a means of scoping the requirements for further analysis and to reduce subsequent efforts. Typically, a qualitative method addresses the sources of errors or lack of knowledge by identifying them as groups, states major assumptions, and characterizes their potential impact on the results.

This may be followed by a quantitative assessment, particularly for parameter uncertainties and input data distributions, in selected areas such as human performance or where sensitivity studies and prior knowledge of data limitations have shown a specific need.

Sensitivity analysis should be used to explore the assumptions in the case of a major impact on results. Typically, these exist where the data or the knowledge base is severely limited and alternative assumptions seem equally plausible. Upper and lower bounds for sparse data should be estimated and used to establish a range of output variation.

As with qualitative uncertainty analysis, sensitivity studies are useful to limit the parameters for which uncertainty must be propagated to those that have a substantive effect on the output of the risk analysis. Use of different models, system success criteria, input data, and assumptions are particularly good candidates for sensitivity analysis.

Uncertainty Analysis Methods

A variety of analytical and numerical methods are available to the ana**l** for the propagation of uncertainty measures through the assessment. **T** most popular methods are analytical and numerical integration, listed h as complementary cumulative distribution functions (CCDF), and discr probability distributions (DPD).

A. Initiating Events
    1. Classical estimation by distribution functions
    2. Bayesian techniques

B. Plant Systems Analysis
    1. Fault tree analysis
        a. Classical estimation by distribution function
        b. Method of moments
    2. Event tree analysis
        a. Discrete probability distribution (DPD) arithmetic

C. Consequence Analysis
    1. Complementary cumulative distribution function (CCDF)
    2. Sampling statistics (Monte Carlo)

COMMUNICATING QUANTIFIED RISK ASSESSMENT RESULTS

Objective

The overall objective of a risk assessment must be clearly reflected in communication of results to others. Risk assessments are often comp enough that a context for the results must be created throughout the rep so that the quantified risk is rendered comprehensible. One comm method is to provide a perspective of comparison with other known acceptable risks.

Relative Risks

A sample of complementary cumulative distribution functions was use the Reactor Safety Study (WASH-1400) to show the relative risk impo by nuclear plant operation in comparison to hazards imposed by fi explosions, air crashes, and dam failures. Figure 13-1 shows that relative risk imposed by 100 nuclear reactors is a small fraction of the imposed by the sum of these other events in terms of early fatalities.

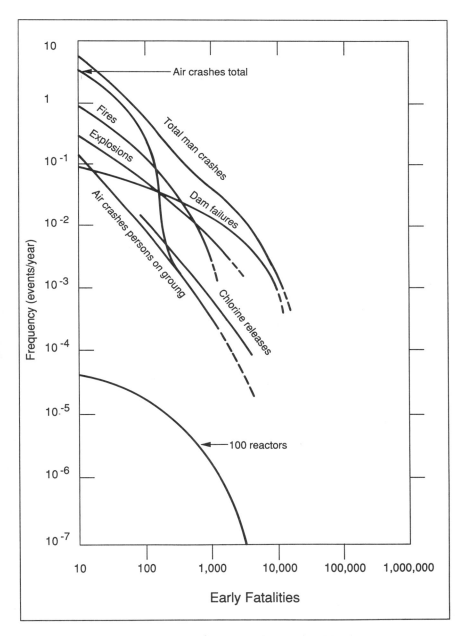

Figure 13-1   Frequency of man-caused events involving fatalities.

229

Risk Contours

Figure 13-2[4] shows how each point on a CCDF curve is generated from the cumulative probability that a release effect or damage is greater than a certain quantity. When these risk curves are shown with the uncertainty included, a set of curves may be developed as shown in Figure 13-3. The set of curves shows the effect of using a probability distribution at each

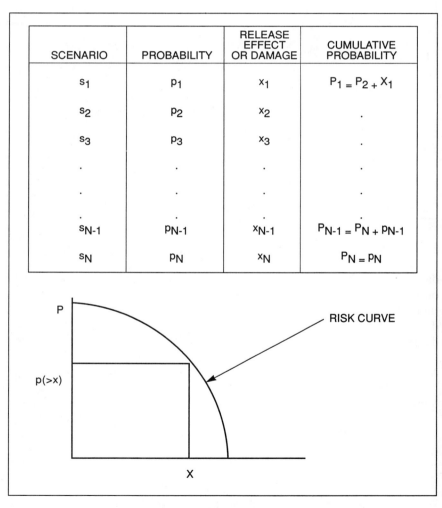

SCENARIO	PROBABILITY	RELEASE EFFECT OR DAMAGE	CUMULATIVE PROBABILITY
s_1	p_1	x_1	$P_1 = P_2 + X_1$
s_2	p_2	x_2	.
s_3	p_3	x_3	.
.	.	.	.
.	.	.	.
.	.	.	.
s_{N-1}	P_{N-1}	x_{N-1}	$P_{N-1} = P_N + P_{N-1}$
s_N	P_N	x_N	$P_N = P_N$

Figure 13-2   Graphical presentation of risk.

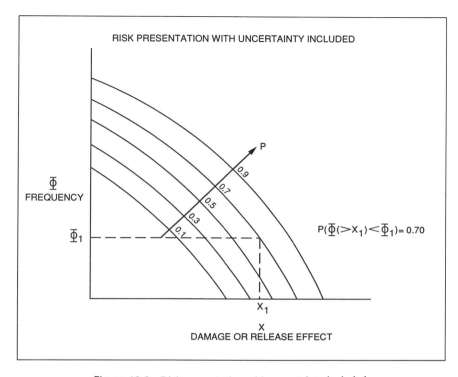

Figure 13-3   Risk presentation with uncertainty included.

damage value of x instead of a point value mean. The example point shown on the curve by the rectangular coordinates is interpreted as saying that there is a 70 percent probability that the frequency of occurrence of damage greater than x_1 is less than $\overline{\Phi}_1$. This expresses a degree of confidence in that particular result.

Acceptability of Risk

Many studies have been made in the last decade about the societal acceptance of risk. These studies generally show that risks are viewed as acceptable if they are natural, familiar, and voluntary or are the result of activities that are viewed as necessary. Whether or not these risks are numerically equivalent or are quite different is often not an issue. Part of the reason for this is that human beings are often illogical and have preferences that are totally subjective. Hence a risk may be shown to be acceptable if the activity has a benefit that is believed to be more than compensating for the risk, or if the risk has been reduced to an acceptable minimum by a defense in depth and mitigation of consequences.

Rather than attempting to characterize risk in any absolute terms, it is perhaps more important to state the relative reduction in risk posed by certain improvements or to show the effect of alternative design choices in minimizing the risk to the public. The cost effectiveness and effect of diminishing returns for additional safety measures is persuasive in showing the alternate uses of funds for the greatest overall benefit.

APPLICATION EXAMPLE

Background

A simple example taken from Henley and Kumamoto[5] is used to illustrate the benefit of quantified risk assessment. The example is based on one type of risk analysis called cause-consequence diagrams, which were invented in Denmark at the RISO Laboratories. For other examples of quantified risk assessment, refer to other chapters, such as Chapter 7 for failure modes and effects analysis, Chapter 9 for fault trees, Chapter 8 for HA-ZOP, and Chapter 14 for human reliability analysis.

Cause-consequence diagrams start with an initiating or critical event that is chosen as a logical starting point for a series of possible event sequences. A typical critical event might be

Breach of a pressure boundary
Initiation of a batch reaction or startup procedure
A transient that activates a safety system
A phase-to-phase or phase-to-ground electrical short.

A cause-consequence analysis traces the sequence of events from the initial critical event through to a variety of possible conclusions, branching at each decision point toward success or failure, much like an event tree. A fire started within an electrical junction box may or may not propagate to involve an increasing scope of the plant, depending on how soon it is detected, how quickly the circuit is opened by safety devices, or when possible fuel sources are isolated or neutralized by fire protection or extinguishing systems.

The construction of the cause-consequence diagram allows the analyst to apply simple estimates of probability and consequence to each event and to construct what is known as a Farmer risk assessment curve from the analytical results. Also, since the sum of the event risks are considered representative for the initiating critical event, the total risk posed by that critical event is obtained as well.

Application Example

Figure 13-4 shows the cause-consequence diagram, which is initiated in the lower right-hand corner by a motor failure. The ensuing fire could just as easily have been initiated by a leak of flammable material, because ignition can occur from any number of stationary as well as portable sources.

The top event of the initiating fault tree, "motor overheats," has a probability of .088 in six months and is labeled event P_0. The probability that the motor overheating results in a fire is estimated at 1 in 50, or .02. At the first branch point in Figure 13-4, the first consequence that could result from the event is labeled C_0, and entails a 2-hour delay in the process, in addition to $1,000 of equipment damage. The total consequence, estimating the downtime losses at $1,000 per hour, is

$$C_0 \text{ [consequence]} = \$1,000 + 2*\$1,000 = \$3,000$$

The probability of this consequence occurring is the product of the probability of the initiating event, P_0, and the probability that the motor overheating will not cause a fire, or, $(1-P_1)$, thus:

$$P_0(1 - P_1) = .088 * (1 - .02) = .086$$

Because risk is taken as the product of the probability of event C_0 and its consequence, the risk for this event sequence is

$$C_0 \text{ [risk]} = \$3,000 * .086 = \$258$$

Similarly, the other event sequences can be evaluated with the following parameter values:

EVENT	*PARAMETER VALUE*
P_0	Probability estimated at .088 in 6 months
P_1	Probability of fire result is .02
P_2	Operator failure = 0.1
	Fire extinguisher failure rate = 10^{-4}
	Test period = 1 month, or 730 hours
P_3	Fire extinguisher control failure rate = 10^{-5}
	Fire extinguisher hardware failure rate = 10^{-5}
	Test period = 6 months, or 4,380 hours
P_4	Fire alarm control failure rate = $5*10^{-5}$

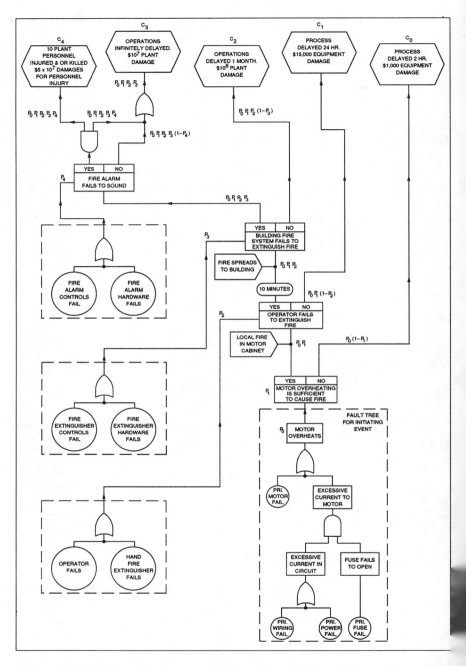

Figure 13-4 Sample system cause-consequence diagram. From Henley and Kumamoto, *Reliability Engineering and Risk Assessment,* 1981, pp. 517, 518. Reprinted by permission of Prentice-Hall, Englewood Cliffs, NJ.

$$\text{Fire alarm hardware failure rate} = 10^{-5}$$
$$\text{Test period} = 3 \text{ months, or } 2,190 \text{ hours}$$

The event probability of event C_1 is obtained from the product of the probability of a local fire in the motor cabinet, $P_0 P_1$, and the success of the local fire extinguisher in putting it out, or $(1-P_2)$. P_2 is estimated assuming that the fire extinguisher fails at a time halfway between monthly tests:

$$Q_2 = P_2 = 0.1 + 10^{-4}*730/2*(1 - 0.1) = 0.133$$

Then the probability of event C_1 is

$$C_1 = P_0 P_1 (1 - P_2) = 0.088*0.02*(1 - 0.133) = 1.53*10^{-3}$$

The consequence of C_1 is the total of $15,000 damage and $24,000 for lost production, or $39,000. The plant risk from this sequence is therefore

$$C_1 \text{ [risk]} = \$39,000 * 1.53*10^{-3} = \$60$$

By an identical process, the remaining event probabilities can be developed and the risk contribution obtained from each event. Table 13-1 shows that the total risk from the motor overheating initial event is about $1,900 per year, and that about 40 percent of the risk stems from potential failure at the hand extinguisher level.

Figure 13-5 is the Farmer risk assessment curve that results from a plot of the Table 13-1 information. Plotted on the curve is a dashed line showing a constant risk line of $300 and implying that events to the left of the line are acceptable, and those to the right of the curve are unacceptable, or at least candidates for further analysis.

Table 13-1 Example Risk Calculation

Event	Consequence	Event Probability	Risk
C_0	$3,000	$8.6*10^{-2}$	$258
C_1	$39,000	$1.53*10^{-3}$	60
C_2	$1.74*10^6$	$2.24*10^{-4}$	391
C_3	$2.0*10^7$	$1.03*10^{-5}$	206
C_4	$5.0*10^7$	$6.69*10^{-7}$	33
			$948 in 6 months or $1,896/year

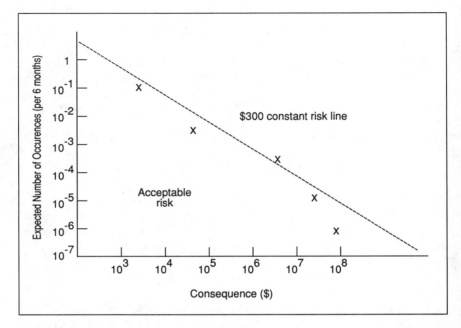

Figure 13-5 Farmer risk assessment curve. From Henley and Kumamoto, *Reliability Engineering and Risk Assessment,* 1981, pp. 517, 518. Reprinted by permission of Prentice-Hall, Englewood Cliffs, NJ.

The risk from event C_3 is from failure of hand extinguishment, since the fire spreads to the building, and most of that from the operator's inability to use the hand fire extinguisher to get the fire out. This explains not only the value of firefighting training, but also the choice of frequent test intervals.

One could also look at the effect of uncertainty in the preceding analysis. Failure rates of fire-protection equipment are somewhat uncertain because of the age of the equipment, undetected or intermittent failures that are not always eliminated by test, and by human error in not restoring the system after maintenance or otherwise disabling it. Even more uncertain is the value of the damage caused by fire, since it makes assumptions about the degree to which the fire can spread in a given time, and the cost of replacing equipment, supports, and structures.

Each parameter and damage value in the foregoing problem could be represented by distributions instead of single values. In addition, since the uncertainty in the cause-consequence model increases with the number of branch points in the sequence, the high consequence end of the curve is almost certainly broader than the low consequence end. All of the foregoing implies that the points on the Farmer curve exist within a "band" of distributions that approximates the dashed line but does not have the same sense as a clear demarcation between what is considered acceptable and what is not. This would prevent any notion that focus on any one event would solve all remaining risk problems.

Summary

This chapter has discussed a few highlights of what a quantified risk assessment should consider from the standpoint of data, modeling accuracy, and interpretation of results through sensitivity studies or uncertainty analysis. A simple example using a cause-consequence diagram was given, along with a qualitative projection of what might result from an uncertainty analysis. The added dimension given by quantified results allows the analyst to focus on the relative importance of parts of the analysis, and possible means for reduction of critical hazards. Numbers, however, are not a substitute for skill and experience in building good representational models, in understanding phenomena, or making insightful links between disparate pieces of information about plants and processes, and their interaction with people.

REFERENCES

1. IEEE Standard 500. Issued 1984. *Reliability Handbook.* New York: Institute of Electrical and Electronics Engineers,
2. Heising, C. D., and Mosleh, A. 1983. A Bayesian Estimation of Core Damage Frequency Incorporating History Data on Precursor Events, Nuclear Safety, *Am Nuc Soc.* 24(4):485–495.
3. *Handbook for Probabilistic Risk Assessment for Nuclear Power Plants,* NUREG/CR-2300, Washington, D.C., 1982.
4. Kaplan, S., and Garrick, B. J. 1981. On the Quantitative Definition of Risk. *Risk Analysis.* 1(1).
5. Henley, E. J., Kumamoto, H. *Reliability Engineering and Risk Assessment,* Prentice-Hall, Englewood Cliffs, NJ, 1981.

14

Calculation of Human Reliability

Robert L. O'Mara

The probabilistic analysis of the chance of a human error occurring in a given sequence of activities is called human reliability analysis (HRA).

Human reliability analysis introduces the methodology and data source usage that can account for human errors in the performance of required operating, maintenance, and test procedures. The end product of an HRA is a set of system success and failure probabilities that reflect the probable effect of human errors on system function. These analyses provide an input to hazardous event sequence quantification and are summarized in the supporting fault trees by task or component. (See Chapter 9 for a complete discussion of event trees and fault trees.)

HISTORY AND BACKGROUND

It has long been recognized that when people are required to operate equipment, the possibility exists of incorrect operation because of human error. The incidence of human error cannot be entirely eliminated, but it can be minimized by the application of proven guidelines for use of the equipment. This activity is called human factors engineering (HFE).

HFE and HRA are relatively young sciences, originating from studies made during World War II, when many people had to learn new technologies in a short period of time. HRA was selected by the Kemeny Commission Report on the Three Mile Island Unit 2 accident and was used as a starting point for nuclear industry improvement, since operator errors were believed to be the principal contributing factors to the accident. Human factors are believed responsible for more than half of the operating event reports currently filed.

THE HUMAN RELIABILITY ANALYSIS PROCESS

Figure 14-1, from the *PRA Procedures Guide*[1], shows in block diagram format the basic process underlying the HRA. It consists of four phases: familiarization, qualitative assessment, quantitative assessment, and incorporation into the risk assessment of which it is a part. The figure is a simplified overview of a process that is essentially an iterative one, with many recursions and interactions between the human reliability analyst and the rest of the risk assessment team. The figure is intended to represent the elements of a complete HRA, but it can be modified to delete steps and to reflect a more appropriate level of detail if only a bounding analysis or crude estimate is desired.

Familiarization

The familiarization phase is intended to acquaint the analyst with the plant design, equipment arrangement, and operation as well as expected failures of the process. Previous system analysis by the risk management team should provide a background source for system failures. In a given scenario or event sequence, human actions are identified that directly affect the critical components of the system in question. Those factors that can alter, negate, or reverse those human actions by changing the input data to the operator, the output performance, or the operator's perception of the event sequence(s) are called performance shaping factors (PSFs).

Qualitative Assessment

The qualitative assessment phase examines the operating procedures by breaking down the individual tasks to elements that have a potential for error. HRA event trees are constructed from the sequence of these operational steps and show the decision points where chosen alternatives result in success or failure of an action. The purpose of the qualitative assessment is to reduce personnel performance specifics to a graphical, symbolic representation of cognitive and physical action that can be used as a framework for the quantitative part of the assessment. The information obtained during the walk-through or "talk-through" of the procedure, done without actually pushing the buttons, should enable the analyst to account for the effect of performance shaping factors. Examples of PSFs are time available, seriousness of the event, and risk of personal injury.

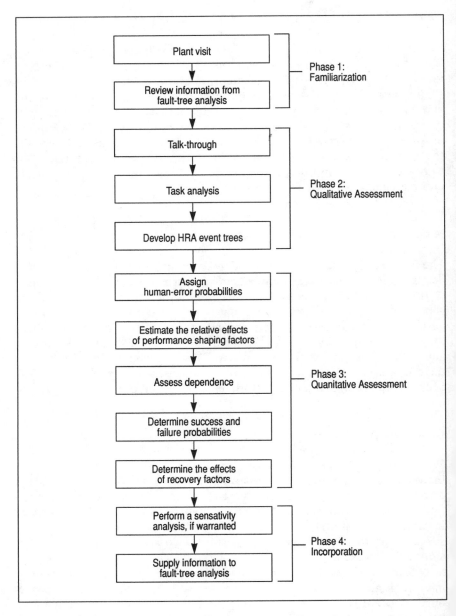

Figure 14-1   Overview of human reliability analysis.

Quantitative Assessment

The quantitative assessment applies human error probabilities (HEPs) to each sequence, or HRA event tree, to obtain the probability of the outcome of the event. The nominal handbook values[2,3] or plant-specific data applicable to these events should be modified by the performance shaping factors found from the earlier talk-through. As described in Chapter 9, the rules for combining event probabilities should also be used to account for dependencies between tasks or operators when more than one is involved. Finally, as shown in the ''Application Example'' section of this chapter, by multiplying the modified probabilities assigned to each branch or limb of the HRA event tree, the probability of success or failure along a particular path in the tree can be determined.

Incorporation of Data

The last phase, incorporation, accounts for the effects of recovery factors arising from operator actions to mitigate or work around the undesired event, identifies the human factor elements that have relatively large or small effects, and summarizes results for the risk assessment team in a format suitable for incorporation into the system event or fault trees.

DOCUMENTS OR INPUT INFORMATION REQUIRED FOR HRA

The system analysis previously performed by the risk assessment team determines which system events or components are essential for system operation and which misoperation can cause its immediate failure. These are called *critical events* or *critical components*. Once these are identified, all defined (or even informally sanctioned) procedures used to perform tasks in these events, or with these components, should be evaluated. This requires submittal of written procedures or preparation of procedure summaries from plant interviews with knowledgeable personnel. The logic or actual mental process used to interpret the control board indications is more essential than the original designer's intent or an engineer's interpretation. It therefore behooves the analyst to try to obtain the informal or routine method by which the control board indications of an event are likely to be read or acted upon, rather than the ''programmed'' response. This will help identify the use of shortcut methods, which often have inherent or embedded errors, misconceptions, or oversimplifications that may be a root cause for human error.

The other set of information required is human error performance data. Figure 14-2 gives a sample selection of such data from the first large scale

Estimated Rates	Activity
10^{-4}	Selection of a key-operated switch rather than a non-key switch (this value does not include the error of decision where the operator misinterprets situation and believes key switch is correct choice).
10^{-3}	Selection of a switch (or pair of switches) dissimilar in shape or location to the desired switch (or pair of switches), assuming no decision error. For example, operator actuates large handled switch rather than small switch.
3×10^{-3}	General human error of commission, e.g., misreading label and therefore selecting wrong switch.
10^{-2}	General human error of omission where there is no display in the control room of the status of the item omitted, e.g., failure to return manually operated test valve to proper configuration after maintenance.
3×10^{-3}	Errors of omission, where the items being omitted are embedded in a procedure rather than at the end as above.
3×10^{-2}	Simple arithmetic errors with self-checking but without repeating the calculation by re-doing it on another piece of paper.
$1/\times$	Given that an operator is reaching for an incorrect switch (or pair of switches), he selects a particular similar appearing switch (or pair of switches), where $\times$ = the number of incorrect switches (or pair of switches) adjacent to the desired switch (or pair of switches). The $1/\times$ applies up to 5 or 6 items. After that point the error rate would be lower because the operator would take more time to search. With up to 5 or 6 items he doesn't expect to be wrong and therefore is more likely to do less deliberate searching.
10^{-1}	Given that an operator is reaching for a wrong motor operated valve MOV switch (or pair of switches), he fails to note from the indicator lamps that the MOV(s) is (are) already in the desired state and merely changes the status of the MOV(s) without recognizing he had selected the wrong switch(es).
~ 1.0	Same as above, except that the state(s) of the incorrect switch(es) is (are) *not* the desired state.
~ 1.0	If an operator fails to operate correctly one of two closely coupled valves or switches in a procedural step, he also fails to correctly operate the other valve.
10^{-1}	Monitor or inspector fails to recognize initial error by operator. Note: With continuing feedback of the error on the annunciator panel, this high error rate would not apply.
10^{-1}	Personnel on different work shift fail to check condition of hardware unless required by check list or written directive.
5×10^{-1}	Monitor fails to detect undesired position of valves, etc., during general walk-around inspections, assuming no check list is used.

Figure 14-2   Sample human error performance data.

.2 − .3	General error rate given very high stress levels where dangerous activities are occurring rapidly.
$2^{(n-1)}\times$	Given severe time stress, as in trying to compensate for an error made in an emergency situation, the initial error rate, $\times$, for an activity doubles for each attempt, n, after a previous incorrect attempt, until the limiting condition of an error rate of 1.0 is reached or until time runs out. This limiting condition corresponds to an individual's becoming completely disorganized or ineffective.
~ 1.0	Operator fails to act correctly in the first 60 seconds after the onset of an extremely high stress condition, e.g., a large LOCA.
9×10^{-1}	Operator fails to act correctly after the first 5 minutes after the onset of an extremely high stress condition.
10^{-1}	Operator fails to act correctly after the first 30 minutes in an extreme stress condition.
10^{-2}	Operator fails to act correctly after the first several hours in a high stress condition.
$\times$	After 7 days after a large LOCA, there is a complete recovery to the normal error rate, $\times$, for any task.

(a) *Modification of these underlying (basic) probabilities were made on the basis of individual factors pertaining to the tasks evaluated.*

(b) *Unless otherwise indicated, estimates of error rates assume no undue time pressures or stresses related to accidents.*

Figure 14-2   (*Continued*)

reactor safety analysis, *WASH*.1400. The collection of such data by Swain and Guttman, 1983[4] is excellent, as is that by Gertman et al., 1988[2]. The use and interpretation of such data should be guided by the handbook text sections of these sources.

GUIDELINES FOR SELECTION OF HRA TEAM LEADER AND MEMBERS

The depth and level of detail intended by the HRA determine the training and experience requirements of the HRA team.

For a complete HRA, the team should include a professionally qualified or trained human factors specialist who has experience in the application to complex systems of such techniques as the Technique for Human Error Rate Prediction (THERP). On the other hand, a bounding analysis or crude estimate can be done by an HRA analyst who is merely familiar with the methodology and the use of HEP data. The remainder of the risk assessment team should include a person or persons knowledgeable about this particular plant or a similar unit's operations and controls. This will enable the event task elements and potentials for error to be fully developed and identified.

At the outset, it should be recognized that a significant item in the quantification of the human reliability event tree is the approximate degree to which its structure represents reality. Because one cause of uncertainty is lack of experience-based data for HEPs, the absolute accuracy of the numerical results may be marginal. When these two factors are considered, it is obvious that one of the most valuable products of an HRA is the relative insight gained as to which operating practices carry the greatest potential risk. These insights are best obtained by using experienced operating personnel who are interested in the analysis, training them in HRA methods, and using them in an effort that is highly focused on sensitive operations or top critical components.

GUIDELINE FOR PERFORMANCE OF AN HRA

The object of an HRA is to "treat the relevant human actions as components in system operation, and to identify error probabilities that could significantly affect system status"[1].

Familiarization and Qualitative Assessment

Suggested procedural steps for familiarization and assessment are as follows:

1. Obtain event sequences and critical components involved from risk assessment system analysts.
2. Talk through the actual sequence of operations and operator responses to the event sequence, establishing the mental process by moving to the controls in sequence with the operator.
3. Survey the control room layout and location of controls or indicators involved in the event sequence.
4. Note those characteristics of critical controls and displays that aid or hinder the operator in locating, manipulating, or interpreting them.
5. Determine whether the existing layout is logical (i.e., promotes correct association of indications or controls, eliminates confusion, and aids interpretation).
6. Determine background material for performance shaping factors by noting timing of steps, stress level imposed by likely alarms and

process conditions, and difficulty of event diagnosis from feedback information.

7. Document results by copious notes on the written procedures and supplement results by photographs or sketches.
8. Break down the event responses into procedure task elements that describe each mental and physical step. (This information is shown schematically in Figure 14-3 which identifies the equipment on which the action is performed, the steps required of the operator, the location of controls and displays, and the potential operator errors.)

HRA Event Trees

Development of an HRA event tree should progress as follows:

1. Construct an HRA event tree from the sequence of operator actions or steps in the task analysis. Chronologically, the tree proceeds from top to bottom. Each success branch of the tree is drawn to the left, and the corresponding failure branch for the same step to the right. An example of such an event tree is shown in Figure 14-4. For illustration purposes only, this tree is drawn with success branches represented by solid lines, and failure branches with dashed lines. Usually, both are solid lines. The HRA event tree also summarizes the total success or failure probability by the symbols S or F, respectively.

   Capital letters represent failure paths, and lowercase letters represent success paths. These are all conditional probabilities, but they are not explicitly labeled as such. For example, the leftmost branch labeled b actually means b conditional upon a, or b|a.

   Commonly, as in Figure 14-4, HRA event trees are developed along a complete success path only, since all the information for a complete tree is contained in the diagram, and the additional branches that would otherwise be shown would be a duplication.

2. Apply nominal human error probability values obtained from Figure 14-2 or one of the reference handbooks[2,3] to each of the HRA event tree branches. Success and failure are complementary (i.e., adding up to 1.0), so the failure probabilities establish the success probabilities, in turn. Some judgment is needed to be able to use the tables of data, since one has to match steps or task elements with the table entry descriptions. The point of breaking the tasks down into smaller incremental steps now becomes clear; the smaller the step, the easier it is to recognize the similarity between the data and the event step

Step	Equipment	Action	Indication	Location	Notes	Errors
D.2	RCS pressure	Monitor		CB4		1. Omission (all) 2. Reading
	RCS temperature heater switches	Monitor Maintain pressure and temperature	Within curve on chart	CB4 CB4		Reading Reading
D.4	4 HPI MOVs	Override and throttle		CP16, CP18	ESF	1. Omission (all) 2. Selection (1)
		Initiate cooldown	Procedure 12			Omission
D.7.3	CV-7621, 22, 37, 38 (room-purge dampers)	Secure	Close switches	Ventilation room		1. Omission (all) 2. Selection (each)
D.7.4	Decay-heat pumps	Verify on	Indicator lamps	CP16, CP18	ESF	1. Omission (for MOVs too) 2. Selection 3. Interpretation
	MOV-1400, 1401	Verify open	Indicator lamps	CP16, CP18	ESF	1. Selection 2. Interpretation
D.9	Borated-water storage tank	Monitor level	> 6 feet	CP14		1. Omission 2. Reading
	MOV-1414, 1415	Verify open	Indicator lamps	CP16, CP18	ESF	1. Selection 2. Interpretation
	MOV-1405, 1406	Open	MOV switches	CP16, CP18	ESF	1. Selection 2. Reversal
	MOV-1407, 1408	Close	MOV switches	CP16, CP18	ESF	1. Selection 2. Reversal
	MOV-1616, 1617	Close	MOV switches	CP16, CP18	ESF	1. Selection 2. Reversal

Figure 14-3 Job task analysis sample.

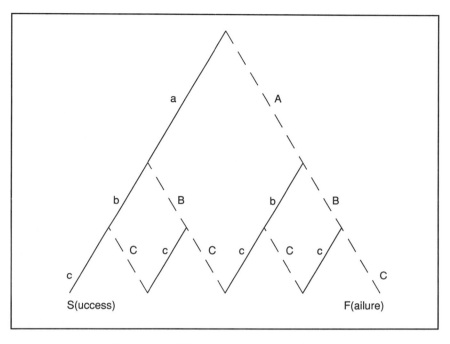

Figure 14-4   HRA event tree example diagram.

addressed. Otherwise, previous experience in the use of such data is needed.

3. Next, estimate the relative effect of performance shaping factors (PSF) acting on the equipment environment, or the individual performing the action. High stress or severe time constraints tend to reduce the probability of success or increase the likelihood of failure. The upper bound of the human error probability value, or a probability that is closer to that upper bound, should be selected if the plant situation is likely to result in error. If the reverse is true, then the lower bound or a lower value than nominal could be used. Each decision point or branch of the HRA event tree should be considered separately for the effect of PSF. If there is a generic or global effect, such as from operator experience level, or stress, then the PSF should be considered for most, if not all, of the event elements.

4. Fourth, the HRA event tree must be examined for dependence between event elements, that is, those factors that affect the conditional probabilities. A very simple example follows that illustrates such a situation:

Two operators are used in sequence to check a reading before deciding to switch the suction point for some pumps. One is an auxiliary operator, the other is a senior qualified operator.

If the auxiliary operator initiates the task, and the results are thoroughly checked by the senior operator, the probability of a correct decision is judged to be high.

If the senior operator goes first, however, it is likely that the auxiliary operator will not question his or her decison, but it is probable that the decision will be a correct one anyway, because of the senior operator's greater experience.

If the auxiliary operator has a good relationship and reputation with the senior operator (e.g., the auxiliary operator's decisions typically receive only a cursory review from the senior operator), then only a lesser experience may be applied to the decision, particularly if the senior operator is busy or preoccupied.

All three of these example cases show elements of dependence, which are tied to mind-set, experience level, and order of the participants in the process. They also show how event- and plant-specific such event analyses are. The reader is warned that dependence may or may not be symmetric in that the degree of dependence may be different for success and failure.

The computational rules for handling dependence are found in Chapter 9. It is more important, however, to be able to recognize dependence than it is to be able to calculate adjustments to probability to account for dependence.

5. Determine the success and failure probabilities for the HRA event tree. This will provide the input to the risk analysis fault trees. It is done by multiplying the series of probability factors together for a complete path or set of paths on the event tree.

    (a) For the complete success path in Figure 14-4, the probability values of the elements a, b, c, d, e, f are multiplied together and the product is labeled S.

    (b) For the total of the failure paths, calculate the sum of the products of all of the success paths down to each failure branch point and each failure branch probability:

$F = F1 + F2 \ldots F6$, where $F2 = a*B$, and $F3 = a*b*C$, etc.

The reason for doing each step in turn is that improvements made because of recovery factors may be applicable to only certain steps.

6. Recovery factors, or partial new success paths, are applied as a

correction to the HRA event tree for the purpose of obtaining an improvement in the success probability of the tree for operator recovery actions. (See Figure 14-5.) These may be accounted for by adding a success branch to the tail of one of the failure branches that leads back to the success side of the tree, or by effectively eliminating the possibility of an error.

7. Finally, check the sensitivity of the results to the error factors inherent in the HEP data[2,3,4] to see whether the outcome of the HRA event tree is heavily influenced by them. Data can be obtained from Gertman et al. 1988[2] and from Swain and Guttman 1983[3] and evaluated with the methods recently applied by Samanta et al. 1989[4].

APPLICATION EXAMPLE

The following is an example of the application of human reliability analysis in support of an accident event sequence analysis for a reactor.

Operator action for controlled primary depressurization is modeled in the small pipe rupture and instrument tube rupture event trees. This operator action, shown in Figure 14-6, is addressed only for those sequences where the following have already been successful: reactor trip, safety injection actuation signal, high pressure safety injection, and auxiliary feed actuation plus secondary cooling. The operator would perform this action in the following cooldown and emergency depressurization procedure.

This procedure requires that the operator reduce secondary pressure to reduce primary system temperature, and reduce primary system pressure by throttling safety injection. By performing these actions, the operator would reduce the break flow and would also reduce the pressure buildup in the containment structure. In addition, this model considers an optional task for the operator to override containment quench spray actuation to conserve the refilling water storage tank (RWST) inventory. The net result of these actions would be that high pressure recirculation cooling would not be required, and, thus, possible recirculation failures would be avoided.

The estimated time frame available to the operator should be sufficient to allow recognition of the event and performance of appropriate corrective actions. In addition, if quench spray is actuated and the operator terminates it within 30 minutes, sufficient storage tank inventory would still be available for long term primary inventory control.

This event has the following tasks, which, with the exception of optional task 3, are depicted in Figure 14-5:

1. Reduce secondary pressure to reduce primary temperature.
   (a) Open steam dumps to the condenser (if available).
   (b) Open relief valve operated by steam generator power (PORV).

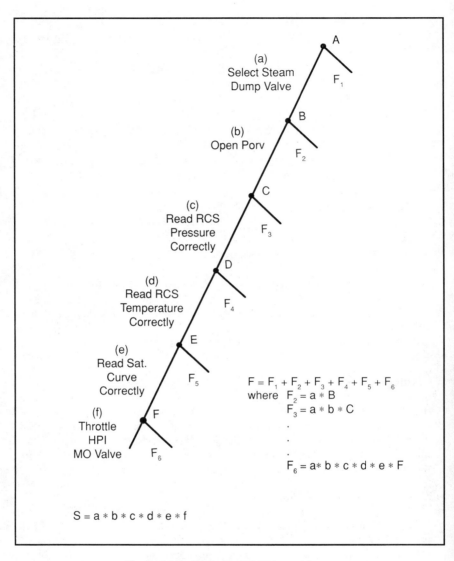

Figure 14-5 Human reliability event tree.

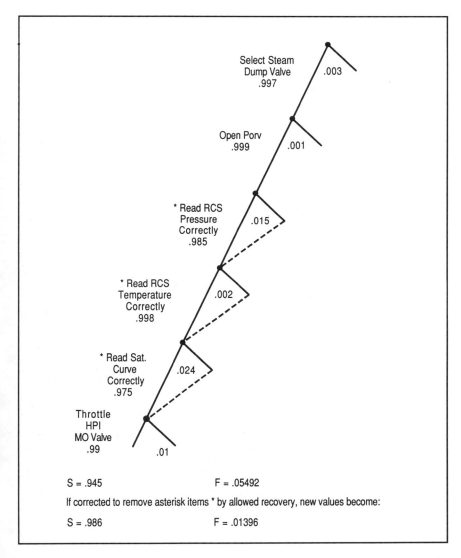

Figure 14-6 Application example. Event OA—2 Operator action. Controlled primary depressurization.

2. Reduce primary pressure to reduce break flow.
   (a) Throttle safety injection while maintaining primary subcooled.
3. If needed, override quench spray actuation to conserve storage tank inventory.

Studies of the flow rates and storage tank inventory history showed the time frame available to the operator to be more than 30 minutes. This relatively long period allows the operator time to recover any procedural errors that may have occurred. Several alternate methods exist that provide the operator with means to perform a controlled depressurization should any specific equipment fail.

Because a long time is available to the operator, the dominant human error would be a cognitive one. Several opportunities are given to redress an incorrect assessment of reactor conditions, since the conditions are changing and present several new cases during the period. In Figure 14-5, this is shown to be equivalent to virtual elimination of the third, fourth, and fifth steps, which involve correct reading of the pressure, temperature, and saturation curve positions. The revised probabilities for success and failure are shown at the bottom of the figure and allow the human error rate to be estimated.

Because several equipment paths are available to conduct the depressurization, human error probability would dominate the equipment failure probabilities for the total system.

REFERENCES

1. *PRA Procedures Guide: A Guide to the Performance of Probabilistic Risk Assessments for Nuclear Power Plants.* January 1983 NUREG/CR-2300.
2. Gertman, D. I., Gilbert, B. G., Gilmore, W. E. et al. June 1988. *Nuclear computerized library for assessing reactor reliability.* Washington, D.C.: NUREG/CR-4639, Vol 5, Data Manual, Part 4: Summary Aggregations.
3. Swain, A. D., and Guttmann, H. E. October 1983. *Handbook of human reliability analysis with emphasis on nuclear power plant applications.* NUREG/CR-1278. Washington, D.C.
4. Samanta, P. K., Wong, S., Haber, S., et al. April 1989. *Risk sensitivity to human error.* NUREG/CR-5319. Washington, D.C.
5. Wechsler, D. *Range of Human Capacities.* Baltimore, MD: Williams and Wilkins.

15

Training for Industrial Facilities

Robert W. Myers and Dixie J. Finley

Over the past several decades, industrial facilities have undergone rapid technological change. As an example, a chemical production facility today is much more efficient and generally more highly automated than a similar facility was 20 years ago. This level of automation, however, has not eliminated human participation. Though much process control is now performed by computer-controlled systems or subsystems, personnel are still needed to operate, maintain, adjust, and calibrate equipment.

Management of industrial facilities includes the responsibility for ensuring employee and public safety. Training of employees is a major element of ensuring reliable and safe human performance of operation and maintenance activities. Training further provides the ability to implement new and improved methods that are developed as a result of new technology or industrial operating experience.

This chapter describes the essential considerations necessary to develop an ongoing training program.

TRAINING PROGRAM

Management Support

An important aspect of a training program that is sometimes overlooked during its development is adequate support by plant management. This is an essential element for success of the training program, and ultimately for the success of the employee. Written corporate and plant training goals should be established. Training programs need adequate funding and staffing to be properly maintained and conducted. Policies and procedures should be established to guide personnel in conducting training activities and in maintaining the integrity of the training program.

Scheduling

Scheduling keeps the training program functioning effectively. A centralized training program is most efficient, but if there are several training programs at your site, scheduling should be coordinated between the various departments.

Scheduling should be completed for a year in advance so that work schedules will not be negatively affected. Schedules should be flexible enough to allow for changes resulting from plant operations. Initial training can be offered at pre-set dates and times each week to facilitate the overall process. A database can be used to list names and type of training required.

Facilities and Equipment

A dedicated facility precludes the effort of coordinating locations with other departments and setting up inside the room before each class. A physical area dedicated to training is needed that has several classrooms available for training to take place concurrently. Operations training, maintenance, health and safety, and fire brigade training, for instance, may be scheduled at the same time. Adequately sized rooms, video equipment, slide projectors, screens, overhead projectors, and appropriate furniture are also basic requirements. If a training department is not designated at the facility, trailers can be converted for this purpose. Many industrial facilities have found that the modular trailer unit provides the desired flexibility for future site expansion projects. Modular trailers can be relocated to suit specific site needs. There should be an adequate number of demonstration models for student use for hands-on training.

Procedures

Written training procedures provide a guideline for the training staff. Procedures describe the purpose and scope of the program. They provide the appropriate references and trainer qualifications, and they outline courses and their content. They also cover any other subjects necessary for the program to function adequately, such as exemptions from training, criteria for successful completion of a course, and evaluation of the system. A matrix designates the modules needed for each job position. The matrix (to be discussed later) serves as documentation of training offered per job title. Not all technical staff members need the same training modules. Their needs depend on the information required to perform a specific job.

Learning Objectives

Learning objectives are written to define the expected performance of the trainee. They also define the material to be developed for the lesson plan. They can be developed from an analysis of what the employee needs to know to satisfactorily perform a job. The learning objectives are always presented to the trainee at the beginning of the class. The trainee then knows what actions he/she is expected to perform upon completion of the lesson. Any limiting conditions for performance are included in the written objectives, for example, if a system must be in standby for an action to take place, or if an emergency condition must exist. The learning objectives can be cross-referenced in the lesson plan and should be reviewed at the end of each unit of instruction to enable the trainee to focus again on his or her expected performance.

Lesson Plans

Lesson plans and other training materials used during classroom, laboratory, individualized instruction, or on-the-job training should be developed systematically and should be based on the learning objectives. Information to support each objective is grouped together. The information for each objective is then arranged to support the objective in the most logical format possible. The learning objectives can be cross-referenced in the lesson plan. The lesson plan should have the same flow or order as the lesson objectives and should be reviewed at the end of each unit of instruction. Much more information can be gathered than is actually used in the module.

Graphics should be developed to support the lesson plan. These can be an important and effective tool in keeping the student's attention and in aiding his or her learning capability. Many computer software programs that have strong graphic capabilities are now on the market (e.g., Freelance, Harvard Graphics, Windows). Graphics can be in the form of slide programs or text. It is important for the instructor to have a manual written specifically for classroom instruction so that the trainees experience smooth transitions from one lesson segment to the next. Graphics are also used to emphasize important concepts and to guide the instructor as he or she gives specific examples or presents the material.

Evaluation

Trainee performance should be evaluated consistently and reliably. Tests and performance evaluations should be based on the learning objectives.

Mastery of learning objectives can be evaluated using any combination of written and oral examinations. Based on examination results, remedial training and reexamination can ensure that performance standards are met satisfactorily. Evaluation or on-the-job training addresses essential knowledge and skills prior to job and task qualification. If the trainee does not achieve a passing score during training, remedial training and further evaluation are needed. Evaluations should be conducted only by designated individuals who are totally familiar with the job function and who can provide consistent and effective training and guidance.

Job Analysis

Information obtained from a job analysis is used to develop learning objectives and the corresponding lesson plans for training modules. An analysis is performed for each job to determine actions performed in that function. The analysis can be accomplished by subject experts who complete a desk-top evaluation, or by an interview of the personnel who perform those tasks on a daily basis (or those who perform the function as a part of emergency response).

A list is then made of all the duties and subsequent tasks associated with the job. For the list of tasks, a skills and knowledge level can be determined. All of this information should be reviewed by someone knowledgeable in the discipline who is not part of the analysis. Reviewing the knowledge and skills needed to perform the tasks will suggest the proper method of training needed, such as classroom, lecture, laboratory practicum, on-the-job training, or individualized instruction. Some level of skills and knowledge may be required of an employee prior to employment. Some employees may be exempted from training due to their education or prior experience; these exemptions should be made only after appropriate evaluation.

TRAINING METHODS

Several types of training have been shown to be effective in the training of adults in work-related activities. On-the-job training, for example, predates the industrial revolution, and apprentice programs have historically been used to foster mastery of job skills.

Formats that have evolved include the following:

- Classroom lecture
- Laboratory practicum
- On-the-job training

- Individual instruction
  Workbook tutorial
  Read and sign assignment
  Computer based instruction
  Computer based tutorial
  Walk-throughs and facility tours
  Simulator training.

Selection of a training format is dependent upon instructional objectives derived from the training program development.

Classroom Lecture

The classroom format is the most familiar in that it is the predominant format used in the public educational setting of U.S. school systems. It is suitable for providing information or knowledge, but it is somewhat limited in teaching action or skills. This format is generally applicable to large groups of students.

Laboratory Practicum

The laboratory practicum is also used in public schools to teach practical skills such as welding, electronics, and chemistry. In industrial settings it is used for maintenance, instrumentation, and related training topics.

On-the-Job Training

On-the-job training is generally designed as a list of tasks that the student learns during performance of the job, first by watching and then by assisting an experienced employee. This initial phase is followed by actual performance under close supervision, followed by a verification of competency to perform the tasks.

Individual Instruction

A brief description of some individual instruction techniques follows:

- Workbook Tutorial
  This format is a self-paced method that is administered with a time constraint in the industrial workplace.
- Read and Sign Assignment
  This format is generally applied when personnel are familiar with or

aware of new management directives, new regulations, and new or revised procedures.

- Computer Based Instruction
  This format is most commonly a PC related application, although it can be on other computer systems. This format provides the student with training information by scrolling text or graphics in user-interactive sequences.
- Computer Based Tutorial
  This format is very similar to computer based instruction. It differs by having a student performance evaluation mechanism included.
- Walk-throughs and Facility Tours
  This format is usually used for familiarization with facilities that are not used on a daily or other regular basis. An example of this is familiarization with emergency response facilities prior to drills or response.

Simulator Training

The scope of simulator training can vary from a PC based interactive program to a mockup of a full control room (common in the nuclear industry). This format, irrespective of scope, is used in training operators in system or unit process evolutions.

TYPES OF PROGRAMS

Initial Training

Initial training usually covers a large scope of material. Health and safety, emergency response, administrative procedures, and detailed technical information are covered. Initial training gives the employee the information needed to achieve the knowledge or skills that will enable him or her to begin job responsibilities. For operations, maintenance, and other technical jobs, initial training usually includes a plant specific period of on-the-job training.

Continuing Training

Continuing training is needed to keep technical staff informed of any new process, plant modification, procedural revision, new regulatory requirement, or change in industry standards or recommendations. Continuing training also serves as an annual refresher course for technical skills, health and safety, and other plant specific information.

Reviewing of lessons learned from industry or in-house operating experience is a form of continuing training. It is a valuable tool for improving operations and maintaining the plant in the safest condition for employees and the public. Staff are trained by looking at events in detail: causes, sequences of events, and consequences. Any early indications of a continuing problem are important to note, as are mitigation and abatement efforts. Corrective actions are studied and applied when appropriate.

DOCUMENTATION OF TRAINING

Documentation of training demands special treatment, because good communication and record-keeping are essential for any training program to be successful. If the training has been performed, but no records have been kept, it is as if no training took place. Documentation can be shown to a regulator or another interested party to demonstrate that you have a functional program, and provides management the confident assurance of a safe, efficient workplace.

Record-keeping is just one aspect of documentation. The whole program should be written and updated at specific intervals. This written program includes, but is not limited, to the following:

1. Overview of program
2. Approval sheet
3. Learning objectives
4. Equipment and materials needed sheet
5. Lesson plans and student handouts for each course
6. Revisions to lesson plans noted with date or revision and reason for revision (e.g., procedure change, lesson learned, equipment change)
7. Evaluation with answer sheet, if appropriate
8. Qualifications of instructors
9. Records of attendance for each training course.

The records for each course may include the trainee's name, date, course number, test score, required classes, exemptions and, if continued training, the date next training is required.

Information can also be kept on position qualifications and personnel qualifications. A distinction can be included as to method of training: on-the-job, laboratory, classroom, or other. If an employee misses training, his or her name should be flagged.

A matrix of courses required for each job title should be developed as a visual representation of the training requirements of the program. The matrix can then be updated as necessary.

A computer word processor and data base are quick and efficient ways to keep a written program and update it as needed. For record-keeping, use a data base that gives some flexibility (for changes in program structure) and the ability to search by subject (e.g., to call up all names that have completed respirator training during the month of June).

Hard copies of records should be kept for a lesser period of time, and on disc or microfiche or another archival retrieval method for a longer period of time. It is recommended, however, that the record retention period be clearly defined and stated within the training program materials or training procedures.

REQUIRED TRAINING MATRIX

Management of industrial facilities includes the responsibility for assurance of an adequately trained work force. Safe, efficient, and reliable operation of industrial processes is the end result of appropriately implemented training.

Definition of training requirements by use of a matrix developed from the job task analysis is the most reliable way of assuring that the work force is adequately trained. As depicted in Figure 15-1, training topics or groupings are established. Each position is then reviewed with respect to the tasks to be performed against the topic listing. The resulting array provides definition of training to be required of each position. From this array, initial training, prerequisite knowledge, and refresher training requirements may then be established.

Training Process Summary

To begin the training process, you must identify your needs for each of the functional departments (e.g. operations, maintenance, engineering, technical support). The training program is put into effect by a continual process. Training materials are developed or modified after looking at the job duties and tasks for each position and the knowledge and skills required for those tasks. Training methods are picked and personnel in need of training identified. Then, training is scheduled, conducted, and thoroughly documented by qualified instructors. An evaluation of trainee performance is included in the documentation. At appropriate intervals an evaluation is conducted for instructor technique and the overall effectiveness of the program. Training materials are continually revised as a result of this evaluation process, as are procedural changes, operating experiences, plant modifications, and new regulatory requirements.

TRAINING TOPICS									
Simplified Example Training Requirements Matrix Position Title or Functional Responsibility	Regulatory Requirements	Process Unit Operations	Welding Procedures	Instrument Calibration	Heat Exchangers	Rotating Equipment	Hotwork Permits	Engineering Control Procedures	Emergency Plan
Facility Manager	X								X
Department Manager	X							X	X
Operator		X			X	X	X		X
Maintenance Worker			X	X	X	X	X		X
Design Engineer	X							X	X

Figure 15-1   Required training matrix.

SUGGESTED READINGS

Davies, I. K. 1981. *Instructional Techniques*. New York: McGraw-Hill.

Knowles, M. S. 1980. *The Modern Practice of Adult Education*. Chicago: Follett.

Popham, W. J., and Baker, E. L. 1970. *Systematic Instruction*. Englewood Cliffs, NJ: Prentice-Hall.

Kletz, T. A. 1985. *An Engineer's View of Human Error*. Institution of Chemical Engineers, U.K.

The Accreditation of Training in the Nuclear Power Industry. INPO 85-002. January 1985, Institute of Nuclear Power Operations. Atlanta, GA.

Fundamentals of Classroom Instruction. 1983. Columbia, MD: General Physics Corporation.

16

Emergency Preparedness

Carolyn C. Burns and Piero M. Armenante

In the 1990s and beyond, the focus of emergency preparedness will be on protecting all environmental media as well as the public health and safety from the potential effects of releases of hazardous substances from industrial facilities.

Before the incident in Bhopal in 1984, emergency planning was the exception rather than the rule for institutions other than the nuclear industry, which was required to respond to the incident at Three Mile Island in 1979. Why did only a few forward-thinking companies in other business sectors develop emergency preparedness programs? Because there was no regulatory impetus for them to do otherwise.

In the wake of the Bhopal tragedy and the accidental chemical release a few months later in Institute, West Virginia, public outrage escalated dramatically. Whether "chemophobia" results from actual or perceived dangers is unclear, but in the public mind, perception *is* reality, and pressure for regulation of the chemical process industries has begun.

The Occupational Safety and Health Administration's (OSHA) Hazard Communication Standard (29 CFR 1910.1200[1]), or Worker Right-to-Know, became fully effective in May 1986. Its purpose is to ensure that employees are aware of hazards in the workplace. Initially limited to industries in Standard Industrial Classification (SIC) Codes 20-39, applicability was extended two years later to essentially all industries with more than one employee.

The public demanded protection, and legislators recognized the necessity of reducing the possibility and magnitude of toxic and hazardous material releases to the environment. The Emergency Planning and Community Right-to-Know Act (EPCRA), Title III of the Superfund Amendments and Reauthorization Act (SARA) of October 1986, was the result[2]. The onus of Title III is on states and local communities to develop plans for

---

NOTE: Prof. Armenante's contribution to this chapter is based on portions of his book, *Contingency Planning for Industrial Emergencies* (Van Nostrand Reinhold, 1991).

responding to hazardous materials emergencies. However, for them to do so requires input from the facilities within their borders as to what chemical releases they may have to face.

More recent federal regulation, that is, OSHA 1910.120 (29 CFR 1910.120[3])—Hazardous Waste Operations and Emergency Response, often referred to as HAZWOPER and promulgated as a result of Section 126 in Title I of SARA, became effective in March 1990. It mandates emergency response and preparedness programs for industry that include required interface activities with off-site agencies, and prompt notification to them of an emergency situation. Unless an employer plans to evacuate employees when a hazardous material release occurs and not try to contain or control it, HAZWOPER's planning and emergency responder training requirements must be met. In any case, employers must have a written emergency action (or evacuation and fire response) plan in place to comply with 29 CFR 1910.38[4], and an employee alarm plan per 29 CFR 1910.165[5] to ensure that employees can be notified of the need to evacuate in an emergency situation.

Some states have even more prescriptive regulations. New Jersey's and Delaware's Toxic Catastrophe Prevention Acts (TCPA)[6] and California's Risk Management and Prevention Program (RMPP)[7] are examples. Industry and local governments (county, municipality, or both) must be prepared to comply with the most stringent requirements applicable to their geographic location.

Nonpreparedness for hazardous materials emergencies has consequences in several arenas. The first may be economic: fines, the cost of compliance, disruption of production and thus, an effect on profitability. These are valid concerns. The potential adverse effects on employee and public health and safety and environmental integrity are more far-reaching. The good neighbor image may also be threatened.

The Chemical Manufacturers Association (CMA) promoted member industry reachout initiatives before federal regulations were promulgated through its Community Awareness and Emergency Response (CAER) program.[8] Louisiana and Texas are one part of the country where CAER was particularly successful. Because of it, much of the groundwork for SARA Title III planning was already in place when the law became effective. In other areas—notably, New Jersey, Illinois, and Pennsylvania—cooperative industry and local government Hazardous Materials Advisory Councils or HMACs are successfully addressing joint concerns.

The planning *process* is the key to success. Relationships among the plant, local government, and the community are often fragile. The process itself provides for interaction among all participants. The planners become aware of one another's strengths and weaknesses, which are factored into the response mechanism to make it workable. The interaction has a positive effect of its own. To circumvent the planning process by agreeing to rework a document prepared by others in a "cookbook" fashion diminishes the importance of those committed to making it work. This chapter will discuss emergency planning for industrial facilities, and planning at the local level.

EMERGENCY PLANNING FOR INDUSTRIAL FACILITIES

Initially, the emergency planning process focuses on analysis of the situation. Members of the planning team use analytical skills to approach the problem. They will, for example

- Analyze the hazards associated with the process and plant as described in previous chapters.
- Assess the resources currently available to control a potential accident.
- Determine the existing command structure at the facility.
- Determine what external resources exist through mutual aid arrangements and at the community level.
- Collect information on applicable codes and regulations.
- Analyze existing plans and assess their validity.

Once this initial work has been completed and relevant material has been collected emphasis switches toward the synthetic part of the planning process consisting of the preparation of the actual plan document. This includes allocation of the resources needed to control accidents, procedures to raise the alarm and assess the severity of the situation, establishment of a chain of command and emergency response structure, and definition of response strategies to protect people, the environment, and property while mitigating an accident. These activities lead to synthesis of the previously accumulated information into a cohesive emergency response organization and structure.

The focus of this section is what should exist in terms of resources and a response organization to control industrial emergencies. The following elements form the backbone of the emergency preparedness program and include the elements prescribed by Section (q) of 29 CFR 1910.120, "Emergency Response to Hazardous Substance Releases"[3]. The subsections that follow are organized somewhat differently from the regulation but contain all of its defined elements, as follows:

- Preemergency planning and off-site coordination
- Personnel roles, lines of authority, training, communication

- Emergency recognition and prevention
- Safe distances and places of refuge
- Evacuation routes and procedures
- Decontamination
- Emergency medical treatment and first aid
- Emergency alerting and response procedures
- Critique of response and follow-up
- Personal protective equipment (PPE) and emergency equipment.

Resources

In order to implement any emergency response action, the appropriate personnel, facilities, equipment, and supplies must be available. Therefore, identification of the resources available for this purpose is an essential part of the planning process. One suggested method is to create a list of those items deemed necessary and compare this "wish list" with what is available. Where needed items are unavailable, a schedule for acquiring them should be established and strictly adhered to.

Facility response teams typically comprise specially trained personnel normally operating the plant itself and therefore familiar with it and its hazards. The plant emergency organization is discussed later in this section. The services of a contractor may augment the facility organization or, in small operations, take its place.

Figure 16-1 presents a sample format for performing a resource assessment. The list is comprehensive; facilities should select only those resources appropriate for their anticipated response activities.

Some of these resources could already be available in local municipalities and at neighboring industrial facilities as well as at the plant. The local fire department, however, may be trained only to cope with the most common emergencies, such as structural fires and rescue, or perhaps for hazardous material transportation accidents. Thus, they may only support the primary response actions that plant personnel must implement in a serious emergency. Local resources will, however, become essential if the emergency spreads beyond the plant boundary.

Neighboring industries can also provide resources and personnel during an emergency. However, plant officials must still assess the resources needed to cope with their identified potential emergency situations. It is the responsibility of plant management to ensure that the appropriate equipment and supplies to respond to their hazard-specific emergencies are available at the plant, independent of external resources.

Whether outside support is provided by local public entities, neighboring industries, or private contractors, it is important that letters of agree-

RESOURCE ASSESSMENT			
RESOURCES	CURRENT	REQUIRED	DEFICIENCY (Acquisition Date)
EMERGENCY CONTROL/ OPERATIONS CENTER			
MEDIA CENTER			
SITE NOTIFICATION SYSTEM			
OFFSITE NOTIFICATION SYSTEM			
COMMUNICATIONS EQUIPMENT			
PERSONAL PROTECTIVE EQUIPMENT			
METEOROLOGICAL EQUIPMENT			
FIREFIGHTING EQUIPMENT			
SPILL CONTROL EQUIPMENT			
MONITORING EQUIPMENT			
MEDICAL/FIRST AID CAPABILITY			
SECURITY AND ACCESS CONTROL EQUIPMENT			
AUXILIARY POWER			
TRAINED EMPLOYEES			

Figure 16-1   Resource assessment.

ment or memoranda of understanding be executed that spell out both what can and cannot be provided as emergency assistance in terms of personnel and equipment. The limitations to such support are as significant as its provision.

Emergency Operations Centers

All but very small plants should establish an emergency operations center (EOC) from which response activities can be directed and coordinated whenever a major emergency is declared or anticipated. (This is not to be confused with the command post established near the scene by the Incident Response Commander under the incident command system, discussed later with emergency response actions).

Upon declaration of an emergency, the EOC will be activated by and comprised of the emergency management staff, including the highest ranking persons in charge of the operation. The EOC should be equipped with adequate communication systems: telephones, radios, and other equipment to allow unhampered communication with the response teams involved in implementing response actions in the field, and with external agencies and response organizations and agencies.

During a fast-moving incident involving the release of a toxic material at a paper mill in a small town in Maine, the phone lines soon became jammed. The required notifications to state and local agencies were made, but in the crush of emergency response activities, no one remembered to contact the media until 3 hours after the start of the incident. False radio and television reports of numerous deaths and injuries had prompted the local hospital to activate—unnecessarily—its emergency plan. Needless to say, additional phone lines, including outgoing only, have since been installed[9].

The EOC should be located where the risk of exposure to accidental releases is minimal. When possible, it should also be located close to routes where it can be reached easily by personnel arriving at the scene. An area near the guard post at the main entrance to the plant can be ideal as long as access to it is strictly controlled. Only a limited and prearranged number of people should be admitted to the EOC when it is in use. This eliminates unnecessary interference and reduces confusion. Security should be in charge of limiting access to the EOC.

The EOC should provide shelter to its occupants against most anticipated accidental releases and especially against infiltration of toxic vapors. A small meteorological station, or at least a wind sock, should be located nearby to monitor wind direction and velocity. Provisions for an alternate EOC should be made, in case the main one is directly affected by the accident or becomes too risky to operate. The location of this alternative site should be chosen carefully so as to minimize the possibility that both EOCs become inoperative.

The communication system should be protected as much as possible against shutdown. The EOC should have an uninterruptible power supply (UPS) or at least backup power for lighting and electric communication system operation.

The EOC should always be ready, or easily set up, for operation. It need not be a single-purpose room; a conference room can be easily adapted for this purpose. If one area cannot be dedicated for the purpose, cabinets containing the equipment and supplies that may be needed during an emergency should be kept sealed in or near the designated area, their contents checked regularly by the Emergency Operations Coordinator, to be opened and set up by the first member of the emergency response organization to arrive when an emergency is announced and the EOC is activated. Recommended equipment and supplies for optimal EOC operation include[8,10]

- Up-to-date copies of the emergency response plan and implementing procedures
- Emergency telephone rosters
- Names, phone numbers, and addresses of external agencies, response organizations, and neighboring industries
- Names, phone numbers, and addresses of employees
- Adequate telephones and lines (some outgoing only)
- Dedicated telephones or hotlines, cellular if possible
- Two-way radio equipment
- Fax machine
- Computer system and cloud dispersion model software (nice, but not required)
- Tape recorder or player (audio, video, or both)
- Clocks (preferably 24-hour)
- Status boards and message boards
- Material safety data sheets (MSDS) for chemicals at the site
- Maps of the plant and surrounding areas
- Transparent plastic sheets to be superimposed on maps to indicate
  Area(s) affected by the accident
  Position and movement of vapor clouds
  Positions of response teams
  Evacuation areas
- Technical and other pertinent reference material.

An effective way to keep track of the several aspects of the emergency, such as personnel who reported for duty, main events, meteorological conditions, and evacuation progress is to use a status board. Magnetic or "stick-on" symbols to be placed on the board are also effective to show the location of relevant points on maps. A status board can be continually updated to track the progress of an incident and response actions for all of the emergency response organization to see at a glance. Typically, these are dry marker, erasable boards with blocks for the time, the incident status, and what response actions have been taken. Sophisticated boards have been developed that can produce a paper copy of the contents at the press of a button or can be electronically transmitted simultaneously to several locations, for example, the media center and local EOC. One individual should be assigned to continually update the board, and organizational procedures should specify the prompt provision of status changes to that person.

Figures 16-2 and 16-3 are photographs of a particularly well equipped EOC showing status boards and the communications area, respectively.

Figure 16-2   EOC Operations Room.

Figure 16-3   EOC Communications Center.

Media Center

A media center is a designated room located on or near plant property where representatives of the various news media would be admitted during an emergency and where press conferences would be held by a plant spokesperson. This center should be located near the plant entrance and be the *only* area accessible to news reporters. Helicopters will probably be circling overhead, with videocameras recording everything for the next newscast, unless the FAA agrees to close the airspace—a doubtful occurrence. However, ground access to the accident scene can be successfully limited. This is important not only so that the facility controls its property, but for the safety of nonplant personnel.

During a large storage tank spill at a refinery, a helicopter was transmitting live shots of the incident. The camera focused in on what the newscaster characterized as the brave employee sloshing around in the spilled, unknown material taking notes on what might have gone wrong. The picture showed a bareheaded man in street clothes, wearing no protective equipment of any sort, standing knee-deep in the diked spill. No plant employee would have been so foolhardy; the alleged "brave employee" was a reporter who had eluded security.

The facility designee responsible for public affairs should ensure that the necessary material is stored at or near the media center. Such material should include a fact sheet on the facility describing the number of employees and annual payroll, taxes paid to the community to reduce the residential tax base, a simple description of the processes, consumer products that are produced either at the facility or ultimately from its operations. It may also include status boards, maps of the plant and surroundings, slide and overhead projectors, microphones, telephones, and any other material that may be useful in the organization of a press conference.

Communication Equipment and Alarm Systems

Communication equipment and alarm systems are essential to make the emergency known both inside and outside the plant, to notify external agencies of the incident, and to coordinate activities among the various groups involved in response operations. Initial notification equipment and procedures are especially important, because their effectiveness determines how rapidly the emergency plan can be activated. Initial notification is also essential, because it provides for early mobilization of outside resources, and because it may be mandated by law if extremely hazardous materials are involved[2,6].

Communication equipment must be available to each function within the response organization to prevent communication breakdowns that could severely cripple implementation of the plan. Procedures to coordinate emergency response actions and to alert the public must also be in place.

Alarm systems to warn employees of a dangerous situation and the need for prompt evacuation are required by OSHA's 29 CFR 1910.165[5].

Horns, Sirens, and Public Address Systems

Audible alarm systems are commonly used in many industrial facilities. Horns and sirens rely on different types and lengths of tones to convey messages. An alarm should not just warn, but also instruct people to perform specific assignments, for example, to go to an assembly point for further instructions or implement protective measures, such as taking shelter in the case of a toxic release. Only if plant personnel and public are familiar with the alarm system and are trained to respond to it can this approach work, however. Horns and sirens can convey only a very few simple messages. Public address systems do more by providing a verbal message. Public address systems may not be as effective as expected, because of the poor quality of audible reception in the presence of external noise. In high noise areas, it is appropriate to install a visual signal (e.g., a flashing light) as well as an audible alarm system. Systems that use belt-worn vibrators as alarm signals are also in use.

A facility situated in a densely populated area identified a potential hazard with severe consequences near the fence line less than 50 feet from the closest house. There was an unused steam whistle on the boiler house. Serious consideration was given to using this very loud signal as a single-purpose alarm to warn employees and neighbors to take immediate shelter if a release was imminent or in progress.

However, whistles and sirens must be used with caution. The management of a mill in Mississippi, with the approval of the community, generously provided high-power sirens to the area so residents would be promptly notified of a dangerous situation. Rather than test the sirens at unusual times when their sounding might be misunderstood, it was decided to activate them at the times previously announced by the firehouse: 6 a.m., noon, and 6 p.m. Being considerably louder than the firehouse whistle, they worked more effectively than alarm clocks but were not appreciated by all of the neighbors. Silent test capability has been added[12].

Telephones

Telephones are often the preferred means of communication for reporting emergencies and for communicating between different areas of the plant. The Emergency Operations Center should be equipped with enough telephone lines to enable all the members of the response teams and functions to communicate effectively. Some lines should be equipped for outgoing-only capability. This can prevent overloading of lines by incoming calls. When several telephones are operating from the same location (e.g., the Emergency Operations Center), it is beneficial to lower the noise level by using sound absorbing materials or by providing telephone cubicles. Adding light signals to the phones is also helpful.

Cellular phones are very useful in emergencies. Whether completely portable or mounted in a vehicle, they can be used to notify response personnel who are away from the plant, by emergency management to direct operations while on the way to the site, and even to coordinate the entire response effort in an emergency, should other means of communication fail.

A large refinery had invested a considerable sum in equipping a large van as a mobile emergency communications center. A serious fire broke out in the tank farm, but when the call went out to bring the van to the scene, it was found to have two flat tires. For a long time, the response effort was directed by the fire brigade chief via the cellular phone in his private vehicle[13].

Hotlines, Dedicated Telephones, and Ring-Down Systems

These systems can play an important role in an emergency by ensuring that a direct communication line is always available between a central location and a large network of users, or among a limited network of users connected together and to a central location. Such systems can be especially useful during an emergency when normal telephone lines are easily saturated.

Portable and Battery Operated Radios

Radios are most effective for communicating with emergency response teams operating in the field. In addition, they can be a backup system in case of telephone communication breakdown. Radios must be checked regularly to ensure that the batteries are charged when the need arises. A frequency dedicated for emergency communications only is preferable to one in daily use by plant personnel. The emergency frequency must also be available to off-site responders or to the site team.

Fax Machines

This equipment is useful during an emergency for quickly receiving and transmitting diagrams, maps, and other information with high visual content as well as for providing hard copy verification of verbal communications. For example, areas that may be affected by a toxic vapor release and require evacuation or sheltering of the public may be easily communicated to the outside agencies in charge of implementing them. Fax machines may also help to save time during an emergency by improving communication efficiency.

Personal Protective Equipment

Personal protective equipment (PPE) is worn by individuals to protect them from a hazard—typically fire or toxic liquids and vapors. The main functions for PPE are to protect personnel from a hazard while performing rescue or accident control operations, to perform maintenance and repair work under hazardous conditions, and to allow personnel to escape from a dangerous area.

Protective clothing for protection against heat radiation or those having high resistance to chemical assault (acid suits) are typically used by personnel involved in firefighting or spill cleanup work. Hard hats offer some degree of protection against falling objects and impacts. However, the most important pieces of protective equipment, in both fire and toxic release accidents, are the self-contained breathing apparatus (SCBA) and the respirator.

SCBA is typically used by personnel performing tasks that require prolonged exposure to a toxic environment such as smoke or toxic vapors. SCBA are commonly used in firefighting and rescue operations such as the rescue of people inside a burning building. In chemical release accidents, personnel can sometimes stop a leak by insolating sections of the plant through isolation valve actuation. If this operation cannot be performed remotely, a team of workers wearing SCBA must get to the valve and perform a manual shutoff. Toxic leaks from ruptured tanks can sometimes be stopped by plugging the leak, an operation that also requires protective equipment. These operations may also require full-body protection against chemicals.

The use of SCBA requires training. It is also important to remember that personnel performing any such containment or control operations must be trained in accordance with the appropriate levels of emergency response mandated by 29 CFR 1910.120 (q)(6)[3]. These requirements are detailed in a later chapter.

The equipment should be stored in strategic locations throughout the plant, in control rooms, EOCs, the firehouse, special plant units, and the emergency supply storage area. A compressor is required for refilling the cylinders. SCBA should be inspected and serviced periodically through a preventive maintenance program.

Respirators are used mainly for escape purposes. One type of respirator is similar to SCBA but comprises a cylinder supplying air for only a very limited period of time (typically 5 minutes), which should enable the person to reach a shelter or escape from the contaminated area. A second type of respirator is an air-purifying device that utilizes a filtering and adsorbing cartridge to provide breathable air to the user. This respirator requires the presence of enough oxygen in the environment for the person to breathe and is effective only if the concentration of toxic compound is from 0.1 to 2 percent[14].

Firefighting Facilities, Equipment, and Supplies

Medium-sized and large plants usually have some type of firefighting capability. Typically, this consists of one or more fire trucks and other emergency vehicles, equipment, and supplies—all stored primarily in a central firehouse location.

Fire pumper trucks are the most important units, being equipped with high capacity centrifugal water pumps. National Fire Protection Association (NFPA) standards provide details on the equipment to be carried on pumpers and on ladder trucks[15,16]. The plant hazard analysis determines the actual firefighting needs.

A firewater distribution system from which water for fire operations can be tapped is common to many industrial facilities. Static water sources such as basins or ponds are also commonly used, provided that a complement of suction hoses is carried on the pumper. Water tank trucks can supply additional water.

Specialized firefighting equipment is often encountered at industrial sites because of the type of fire hazards present. For example, dry chemical units carrying large quantities of dry extinguishing material such as potassium bicarbonate (purple K) may be necessary where water cannot be used as an extinguishing agent[17,18]. Units carrying liquid carbon dioxide can also be used effectively.

Fire extinguishing foam is probably the most frequently used nonaqueous medium. It is the best extinguishing agent available against fires covering a large area, as in the case of an ignited spill of flammable liquid[14,19].

As in all emergency operations, communication equipment such as portable radios or cellular telephones must be available to fire response teams.

Spill and Vapor Release Control Equipment

Few methods are available to control a vapor release after it has occurred. Fixed abatement systems (e.g., water curtains), spraying an absorbent such as water into the dispersed cloud of a soluble vapor, such as ammonia and hydrogen chloride or fluoride are sometimes used[20]. Similarly, water streams from fire hoses can be used to create fogs and to allow response personnel to perform emergency operations such as the rescue of human beings or emergency isolation of damaged equipment. Flammable vapors or gases are not generally absorbed using water as a solvent. However, effective dispersion of gases in air below their flammability point can be achieved using water streams[21].

Special tools may be required to perform some response operations such as plugging a leak or shutting off jammed isolation valves. Repair kits should be available for this purpose[22].

Equipment for stopping and containing a liquid spill is also necessary. Dikes, curbs, and entrenchments are used for containment in the case of fixed storage tanks. Emergency containment systems also can be built, provided that the terrain permits and that earth-moving equipment is available. To limit the leaching of spilled material into the ground and to nearby sensitive areas such as sources of potable water, plastic liners and floating booms are commonly employed. In addition, pumps can be effectively used to transfer the material spilled or the material contained in a damaged vessel to a safe location. Similarly, quick setting foams can be used effectively to create an impermeable barrier to spills. These foams can also be used to temporarily plug a leak[23].

Medical Facilities, Equipment, and Supplies

Most industrial facilities have a medical center staffed by medical personnel (e.g., a nurse or a part-time doctor). This facility should also be equipped to deal with the most likely medical emergencies at the particular site. Information on the emergency treatment of exposure to the hazardous materials present at the plant should be available. A file of MSDSs for this purpose should be maintained in the medical center.

The plant response team personnel should be trained in cardiopulmonary resuscitation and should be equipped with the most common types of rescue equipment. Serious injuries should be treated at the local hospital, however. Therefore, medical transportation equipment (such as an ambulance or a vehicle that can be adapted to the transportation of injured persons) should be available, at least in medium-sized or large industrial facilities.

Monitoring Systems

Vapors or gas leaks can be detected using sensing devices.[24] These detectors can be located at fixed points throughout the plant to provide early warning of toxic releases. Response teams dealing with emergency toxic release emergencies should also be equipped with similar field equipment to monitor the concentration of a released substance at different points.

Meteorological Equipment

During toxic release emergencies, meteorological conditions can greatly affect the impact of the release on vulnerable areas. This is especially true if the magnitude of the release is such that the public may be affected. The most important of these parameters are wind direction and speed. The plant should have one or more meteorological stations where this key information is constantly monitored. The station need not be complex. At one facility, a swiveling toy airplane mounted near the guard gate serves the purpose. Wind direction is indicated by the way the airplane points. The rotational speed of the propeller, connected to monitors in the firehouse and safety office where a continuous record is kept, indicates the speed.

Care should be taken to avoid locations where the presence of buildings and other structures may result in faulty readings. Several tall buildings located in strategic positions could cause building wakes, causing distorted wind directions, which can be confusing to instrumentation.

Similarly, wind socks mounted (as they usually are) on the highest point of the facility for optimum visibility will give indication of the wind direction at that level, which may be very different from that experienced by people on the ground. Personnel evacuating a building need to know in which direction to turn upon exit to avoid a toxic plume. Wind socks mounted at a height of approximately 10 feet near identified points of egress are therefore recommended.

As discussed in previous chapters, computer programs are available to predict the dispersion of pollutants under given emission conditions as a function, primarily, of location and meteorological conditions[25]. These computer programs can be effectively used only if wind direction and speed are known. This makes the determination of these parameters all the more important.

Transportation Equipment

Emergency supplies and equipment must often be transported to or near the scene of the accident. In addition to special emergency vehicles, emergency supplies and equipment can be transported using trucks, vans, and other vehicles used during normal plant operations. However, during an emergency situation it is important that these vehicles be dedicated to emergency operations and their use be coordinated accordingly by the response personnel.

Security and Access Control Equipment

In an emergency, plant security is in charge of directing incoming response teams and resources to appropriate assembly or staging areas. As a consequence, it is likely that traffic flow around the plant may have to be redirected, especially if toxic material has contaminated nearby areas. Equipment such as flares, emergency lighting, road barriers, reflective vests, reflective tape to mark restricted areas, traffic control cones, and similar equipment will be required for this purpose.

ACCIDENT EVALUATION

To make appropriate decisions and take necessary actions in response to an emergency, a continuous appraisal of the situation is required throughout the emergency's several stages. This process begins with the first responder, whomever the person may be that detects an abnormal event. The ensuing accident evaluation process is then performed by the person in charge of coordinating the response action at that particular time, and then, later, by the Emergency Director and his or her staff. However, during the initial stages of the still-undeclared emergency, someone, such as the first responder, must decide whether to initiate the alert procedure. Because people will judge the same situation differently, an accident classification system is necessary.

There are several ways of classifying an event that may develop into an emergency. To eliminate some of the confusion that almost inevitably arises, it is convenient to refer to a set of guidelines which have been previously agreed upon. A practical way to classify the seriousness of an incident and quickly convey this information to other personnel (both inside and outside the plant) is to use Emergency Action Levels (EALs). EALs may be designated by code names or numbers associated with situations of different intensities: The higher the number, the more serious the problem. For most industries, a three-level classification system is adequate. These levels are as follows[6, 10]:

Level 1—Alert. The lowest emergency level, this EAL may be associated with an unusual event that is either under control or can be easily brought under control by plant personnel in the immediate area. Events typically classified at this level are small fires, explosions, or minor releases that can be managed by personnel in the immediate area and have a negligible impact on plant personnel or operations. Depending on the type of incident, external notification may be required, but assistance is not necessary.

Level 2—Site Emergency. This intermediate emergency level is associated with fires, explosions, or toxic releases that affect more than the immediate area but have not spread beyond the plant boundary. Off-site populations are not expected to be affected, although plant evacuation or shutdown may be necessary. This level also implies that plant personnel have not yet or do not immediately expect to be able to control the situation and that external support may be required. Off-site responders such as firefighting, medical, police and spill- or release control personnel should be placed on standby.

Level 3—General Emergency. This is the most critical emergency level and implies that the accident already has or has the potential for spreading beyond plant boundaries. If a toxic release has occurred, the outside population may be affected and, depending on the type of accident, may be instructed to take shelter or evacuate. Medical, firefighting, and other agency personnel may be required. Notification of regulatory agencies is mandatory.

The use of EALs has the advantage of standardizing response to different classes of accidents in terms of the resources mobilized to cope with the emergency. It also improves communications during critical times.

The emergency classification system should be discussed and agreed upon with local officials and also with representatives of other local industries so there is common understanding in case of an emergency. All plant personnel should be aware of the system and its implications, since they will be required to take predetermined actions (including evacuation or taking shelter, if appropriate) when an emergency is declared.

THE EMERGENCY RESPONSE ORGANIZATION

The major objective of contingency planning is the creation of a response organization structure capable of being deployed in the shortest time possible during an emergency. For this purpose, the following questions need to be addressed[10]:

- Who will be in command of emergency operations?
- Will the command structure change as more plant and possibly off-site response personnel reach the accident site?
- How will the command structure evolve if the emergency worsens, more resources are required, or more vulnerable points (including off-site vulnerable areas) become at risk?
- Who will decide what company resources to allocate to mitigate the consequences of the accident?
- Who maintains communication with whom during an emergency?
- Which emergency functions (firefighting, engineering, medical, etc.) should be deployed, when, and how?
- Who will be in charge of each specific emergency response function?
- Where should the command post(s) be located?
- Who decides which protective actions to recommend (evacuation or shelter) to protect the external population?
- Who decides when the emergency is over and authorizes reentry into the area?
- Who is responsible for recovery operations?

In response to these and other questions, a response organization, complete with command structure, should be developed. One method of arriving at this is to first develop a responsibility matrix of emergency organization functions versus the departments or positions that will have primary and support responsibilities for performing them. An example of such a matrix is presented in Figure 16-4.

Initial Response Organization

Since accidents can happen at any time of the day or night the capability to activate the emergency response organization must exist even when the plant may be operating with reduced personnel.

Often, the timely implementation of appropriate initial response actions

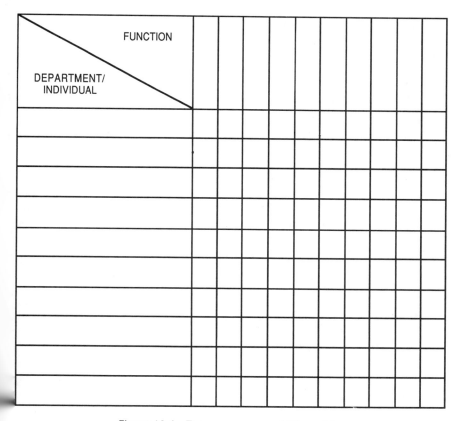

Figure 16-4 Emergency responsibility matrix.

may significantly mitigate the consequences of an accident before it esca-
lates. Therefore, it is essential that only one person be designated to be in
charge of the initial response. Usually, responsibility for coordinating
these emergency response actions is assigned to the shift supervisor until
relieved by higher ranking personnel, such as the plant manager, as this
person arrives at the scene. Initially, then, the shift supervisor assumes the
function of emergency director (or site emergency coordinator) and, as
such, assesses the level of severity of the accident using the classification
system described above; notifies the appropriate personnel, departments
and agencies; and directs the response activities.

Other plant personnel are also assigned the task of covering the other
key functions in the initial emergency response organization until the
predesignated personnel arrive to relieve them. These assignments must
be defined beforehand and not during the accident, and conform as much
as possible to normal job duties. Initially, the post of Response Operations
coordinator is taken by the production or operations manager, who could
also act as the Field Operations Coordinator (or Incident Response Commander). Alternatively, the latter position could be taken by the production supervisor of the unit where the emergency occurs or by the maintenance supervisor, if available. These two coordinators are in charge of
organizing response activities and directing response operations at the
actual emergency scene, respectively. In addition, security personnel will
be in charge of the extremely important initial communication and notification activities, as well as securing access to the facility.

Full Emergency Response Organization

The deployment of the full emergency organization will be required only in
the case of severe emergencies. For example, if a site or general emergency (level 2 or 3) is declared, the Emergency Director will probably
initiate all actions required by the contingency plan, including activation of
the full emergency response organization. An example of a full emergency
organization chart is given in Figure 16-5. The most important functions in
this structure are examined below.

It is important that personnel are assigned to all relevant emergency
functions. For each position multiple assignments should be made so that
either the principal person or one of the deputies can cover that position at
any given time.

The personnel assigned to each of the different functions shown in
Figure 16-5 play key roles in providing advice to and implementing the
decisions made by the Emergency Director. They decide what appropriate
response actions to take: shutting down the plant, fighting fires, evacuating
plant personnel, or recommending that the public be evacuated from

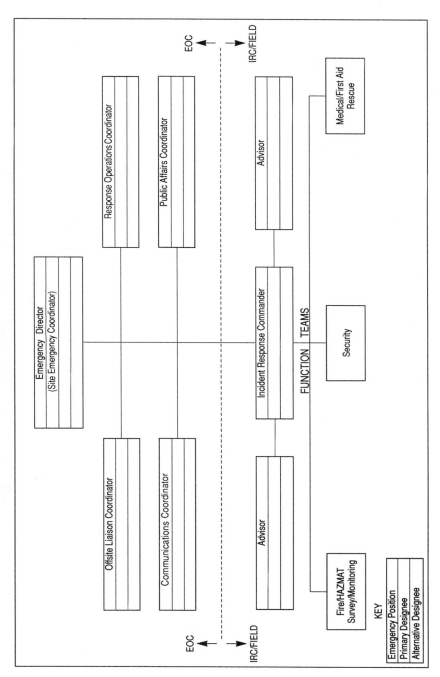

Figure 16-5 Emergency response organization.

certain areas, carrying out emergency repair work, arranging for supplies or equipment, and coordinating actions with local off-site agencies.

Emergency Director (Site Emergency Coordinator)

One person must be in charge of the emergency management efforts. This person is the Emergency Director, whose responsibility includes those actions necessary to bring the emergency under control, the overall supervision of the protective actions recommended for the public, employees, and the environment.

It is good practice to assign the position of Emergency Director as well as any other emergency position to a job title within the plant organization and not to a specific individual. This eliminates the need to constantly update the plan: Positions remain relatively constant; the people filling the positions do not. Names of individuals with their job titles and phone numbers are best listed in an easily updated attachment or appendix.

In addition, the same hierarchical structure used during normal operation should be maintained during an emergency if possible. Accordingly, the plant manager should be assigned the Emergency Director position in the emergency organization structure.

All emergency positions, including the Emergency Director, should have one or more designated deputies for any occasion when the primary designee is not available. Deputies should also be ready to take charge if anyone becomes incapacitated during the emergency.

Response Operations Coordinator

The Response Operations Coordinator operates in the Emergency Operations Center (EOC) and is responsible for correlating the activities of that facility. Therefore, the Response Operations Coordinator performs the following functions[10]:

- Helps the Emergency Director organize and direct emergency activities
- Formulates strategies and advises the Emergency Director on actions to be taken to mitigate consequences of the accident
- Maintains direct communication with the on-scene incident commander
- Coordinates activities aimed at organizing, requesting, and obtaining needed additional personnel and equipment resources.

The Response Operations Coordinator should be very knowledgeable of the plant and its response plan and organization. In small plants or when deemed appropriate, this position may coincide with the Emergency Director function with both positions assigned to the same person. This may also happen during the initial phases of an emergency when the full response organization has not yet been activated.

In other cases, the Emergency Director function may be assumed by the plant manager, mainly because of the responsibilities associated with it. The primary function of the Emergency Director here is that of decision-making and overall guidance, while the bulk of the coordination activities may be performed by the Response Operations coordinator.

Incident Response Commander (Field Operations Coordinator)

The Incident Response Commander or Field Operations Coordinator is the highest ranking officer at the scene of an accident. The command post from which that person directs the emergency response should be located as close as possible to the location where emergency field operations are performed, with due regard for safety (the "safe distance" of 29 CFR 1910.120)[3]. The main responsibilities associated with this position are[10]

- Direction and coordination of all field operations at the scene
- Accident assessment
- Recommendation of on-site protective actions for plant personnel
- Implementation of on-site response actions to bring the emergency under control
- Coordination of these actions with the emergency preparedness coordinator or emergency director in the EOC.

The Incident Response Commander must be very familiar with the facility and have solid technical expertise. This position may be assigned to the fire brigade chief, safety department head, or production manager. During the initial phases of an emergency, the shift supervisor will probably assume this position, at least temporarily.

Emergency Functions

Implementation of an emergency response plan relies on a number of functions that deal with different aspects of the emergency. The most important emergency functions are[10]

- Communications—Ensures that the flow of communications within the on-site response organization, and between this organization and off-site response organizations and agencies is effective and uninterrupted.

- Fire and Rescue—Most facilities have an emergency response team, primarily trained to handle fires and rescue operations and typically composed of personnel from the different plant units or departments. Emergency coverage around the clock is essential. These teams usually have some basic training in the handling of other types of emergencies (such as spills) to control the situation before more specialized teams arrive at the scene of the accident. Team members should also be trained in comprehensive first-aid, search-and-rescue procedures, and emergency equipment handling.
- Special Hazard (HAZMAT Team or Spill Control)—In charge of dealing with any emergency caused by the presence of special hazards such as releases of toxic or hazardous materials. The handling of these emergencies depends on several factors, such as the physical state in which the material is or will be after a loss of containment (solid, liquid, or gas), type of hazardous material (e.g., poisonous gas, explosive, flammable liquid), type of facility and equipment in which the material is handled (e.g., storage tank, reactor, compressor), and type of accident (e.g., vessel rupture, overfilling, fire, toxic vapor release).
- Process/Utilities—Controls the process during the emergency and ensures that the necessary facility units are shut down, as required; responsible for recommending the appropriate procedures to isolate damaged units without introducing new hazards, and for providing resources in terms of personnel and equipment to accomplish this. Also tasked with generating the necessary utilities during the emergency, isolating or recommending emergency isolation procedures to prevent utility distribution to damaged parts of the plant, and activating backup emergency generators upon request. (However, automatic switchover to backup power upon loss of the primary system is strongly recommended.)
- Engineering/Technical Assistance—Provides technical support in connection with the continuity of emergency supplies and services, strategies to isolate damaged process equipment, emergency transfer of materials to safe vessels, and all other process-related emergency operations.
- Environmental and Field Survey—In charge of reducing the impact of an accident on the environment. Develops methodologies to control hazardous material spills and cooperates with emergency response squads to conduct the actual cleanup work during and after an emergency; determines the level of contamination of a site as the result of an accident.
- Medical—Responsible for providing first-aid to the victims of an accident, and arranging for their prompt transportation to a hospital, when

necessary. The coordinator of this function should preestablish contact and cooperate with local hospitals to ensure that the most likely injuries (e.g., burns, toxic inhalation) can be adequately treated at these facilities. He or she also provides information on the nature and properties of the chemicals that could be responsible for possible injuries and suggests the most appropriate emergency treatment of injured or exposed personnel (this information is usually found on the Material Safety Data Sheet [MSDS] of the chemical of concern).

- Security—During the initial phase of an emergency, security personnel at the gate may be in charge of communications within the plant as well as with outside agencies until relieved by the appropriate coordinator. Their responsibility is ensuring that plant security is maintained, making sure that unauthorized persons are not admitted to the plant, and controlling the entry of authorized personnel and the exit of contractor and other personnel.
- Off-site Liaison—Coordinates actions between the on-site response organization and external agencies; initiates and maintains contact with the appropriate external response teams, departments, local representatives, agencies and neighboring industries.
- Public Affairs/Legal Counsel—Since industrial emergencies are likely to attract the interest of the media, it is important that a Public Affairs Function be included in the response organization. The person in this position is responsible for providing news releases (approved by the Emergency Director) and legal counsel during an emergency. In large companies or when deemed appropriate, the Legal Function may constitute a separate function.
- Resources/Supplies—Ensures that the necessary supplies are made available to the emergency response teams; organizes and maintains the staging area where emergency material and equipment is temporarily stored and assembled for rapid deployment.
- Administrative/Clerical—Provides the necessary administrative and clerical support to relieve technical personnel from such responsibilities; keeps chronological records of what is happening during the emergency and prepares reports for the Emergency Director and staff.

Other functions may be added to the response organization if appropriate for a particular site.

The persons in charge of these functions may report to the Emergency Director or to the Incident Response Commander, who in turn is responsible to the Emergency Director às on-scene emergency manager. Each function is staffed by a team, the size and composition of which depends on the task to be performed and the size of the facility. These teams operate according to instructions provided by the Emergency Director's staff and utilize preformatted, written procedures to accomplish their tasks.

EMERGENCY RESPONSE ACTIONS

One important aspect of emergency preparedness is the description of those actions implemented by the various response functions during and after an emergency. The emergency response plan includes some of these concepts; specific details should be incorporated in detailed procedures included as annexes to the plan. These procedures detail the actions to be taken by the various functions or individuals within the organization in response to the potential emergency situations identified in the hazard analysis. For simplicity and user-friendliness, these are best formatted as checklists.

Concept of Operations

An outline of the response sequence should be included in the emergency plan that describes the steps associated with responding to a full-scale emergency, assuming the most severe scenario in the hazard analysis. This description should include a brief description of the following points[10]:

- Warning upon accident discovery (emergency response at the awareness level)
- Accident evaluation and classification
- On-site emergency declaration and response team activation
- Off-site agency notification
- Implementation of on-site response actions (emergency response at the operations level or above)
- Implementation of protective actions
- Coordination of response actions with external resources
- Recovery and plant reentry.

Emergency Response Implementing Procedures

Procedures that enable the emergency response organization to implement the plan may be organized according to the type of hazard; for example, a refinery may have procedures for fire, toxic release, and liquid spill; a pulp and paper mill will have procedures for H_2S and other noncondensible gases, chlorine dioxide, and other substances. For each hazard, a set of procedures is developed, including checklists for each of the function coordinators and teams that detail the steps to follow. These should be specific enough to guide the user through an actual emergency and at the same time be flexible enough to cope with unexpected developments.

Recovery, Plant Reentry and Restoration of Services

One area typically neglected in the emergency plan is post-emergency activities. Specific procedures for recovering from an emergency and reentering a facility must be determined on a case-by-case basis and depend on the type of accident and the severity of the damage. However, guidelines for functional team activities, in both the short and long term following termination, should be included. This preplanning is one of the OSHA-decreed plan segments[3].

Once the critical phase of the emergency is concluded, an inspection team appointed by the emergency director should reenter the damaged area and ensure that it is safe for recovery operations. Personal protective equipment should be worn and portable monitors carried if hazards such as pockets of toxic gases may still be present. In any case, the safety of this team must always be kept in mind.

The impact of an accident may be felt throughout the plant even if only a relatively minor emergency has occurred. For example, damage to a unit may disrupt production in others. A fire in an office may upset sales or delivery schedules. If toxic or flammable materials have been involved in the accident, decontamination work must be carried out before any other restoration activity is attempted. The decontamination methods, materials, and procedures to be employed must be delineated in the plan.

The main objective of the recovery phase is to restore the plant to its initial condition. In addition, one can incorporate lessons learned from the accident and include features that may prevent the occurrence of another such event.

After the emergency is finally concluded, the emergency response plan, the way it was implemented, and its effectiveness in bringing the emergency under control should be reviewed. This should result in improvement. It may be advantageous to assign responsibility for conducting the incident review to the team that developed and is responsible for updating the emergency plan.

TRAINING, EXERCISES, AND PLAN MAINTENANCE

An emergency plan, no matter how carefully prepared, cannot be effective unless accompanied by a training program that includes periodic exercises and drills. The objectives of this facet of emergency preparedness are to[10]

- Familiarize personnel with the content of the plan and its manner of implementation.
- Train new personnel or personnel who move within the plant organization.
- Train specific response personnel in particular duties requiring special skills (e.g., OSHA 1910.120 HAZMAT technician and specialist training[3]).
- Introduce personnel to new equipment, techniques, and concepts of operation.
- Keep personnel informed of changes in the plan or procedures.
- Test the preparedness of response personnel.
- Test the validity, effectiveness, timing, and content of the plan and implementing procedures.
- Test emergency equipment.
- Update and modify the plan on the basis of the experience acquired through exercises and drills.
- Maintain cooperative capability with local response departments, organizations, and jurisdictional agencies.
- Maintain good emergency response capability.

Anyone assigned to a position within the emergency response organization needs initial training followed by periodic refresher courses. Members of off-site emergency response organizations could also benefit from this training. This would strengthen the cooperation among response groups, improve communication procedures, and provide an opportunity for off-site responders to become familiar with areas of the plant where they could be called to assist.

Drills and exercises are vital to emergency preparedness. Both involve enactment, under conditions of a mock scenario, of the implementation of the response actions performed during an emergency. Development and conduct of three types is recommended: tabletops, functional drills, and full-scale exercises. Tabletop drills or exercises are useful for orientation purposes; while gathered around a table, the emergency response organization is presented with a situation to be resolved. Functional drills are more limited in scope than exercises and test a limited aspect of the response capability (e.g., a fire drill). Exercises are more comprehensive and test the entire response organization up to and including communication with off-site response organizations. On-site exercises are strongly recommended before outside agencies are actively involved. Deficiencies that may be discovered during an exercise of the facility plan and procedures should be corrected first.

The plan and procedures must be regularly reviewed and revised, part of the process termed plan maintenance. This is to ensure that they are current and valid, conform to any new regulations, and incorporate lessons learned and modifications that may improve effectiveness.

EMERGENCY PLANNING AT THE LOCAL LEVEL

Although this book is geared to the needs of chemical process facilities, regulatory requirements exist under OSHA and EPA for interaction with local, state and, in some cases, federal agencies[2,3]. For example, SARA Title III—The Emergency Planning and Community Right-to-Know Act of 1986[2] (EPCRA) mandates the establishment of Local Emergency Planning Committees (LEPCs) and the development of hazardous materials emergency response plans by these committees for their communities. For communities to develop these plans, they need to be aware of the hazards presented by facilities within their environs. Facilities that use or produce more than the Threshold Planning Quantity (TPQ) of any of the approximately 360 Extremely Hazardous Substances (EHSs) named in Section 302 of that legislation[2] are required to report this to the LEPC and also to the State Emergency Response Commission (SERC). In addition, facilities subject to the requirements of the law must provide a representative to the LEPC to assist that agency with its planning. For these reasons, an overview of the planning process at the local level is provided in this chapter.

Emergency Management

Effective emergency planning at the local level provides assurance to a community as to how citizens and property will be protected in a disaster. All emergency or disaster plans have the same objectives, specifically to

- Create an ensured level of preparedness for identified potential emergencies.
- Ensure an orderly and timely decision-making process.
- Ensure the availability of necessary services, equipment, supplies, and personnel.
- Ensure a consistent, preplanned response to a given set of circumstances.

The optimum emergency management plan delineates actions that may be required for any hazard, natural or technological[10]. Such broadly applicable functions as direction and control, warning, communications, and public protective actions are as generic to the management of severe weather-driven emergencies as they are to industrial plant accidents. Multihazard, functional plans are thus recommended for smaller government jurisdictions: states, provinces, counties, cities, and towns. These are jointly categorized in this chapter as the local level.

A multihazard emergency operations or contingency plan consists of a basic plan, generic functional annexes, and hazard-specific appendices[26]. The basic plan provides an overview of the jurisdiction's approach to emergency management and its policies as well as its generally applicable capabilities. The generic functional annexes support the basic plan and address the specific activities required in all emergency response. Hazard-specific appendices manage the special problems identified during the hazards analysis process, a prerequisite to the development of any emergency preparedness program. They detail the tasks to be performed by preassigned organizational elements at projected places under specific circumstances, based on plan-defined objectives and a realistic assessment of response capabilities.

Leadership Commitment

Management commitment to the philosophy of emergency preparedness is as vital at the local level as it is for plants and corporations. The motivation for top-level commitment to effective emergency preparedness comes from the concerns of the citizens in the community. They want to know the extent of the hazards to which they may be exposed and how they will be protected against the effects of potential accidents. What warning will they have? How will they know what to do? The community emergency plan answers their questions.

Citizen awareness of the potential threats to safety and health from hazardous materials is escalating, whether the danger is passing through on highways or in industrial facilities located nearby. Local government officials have the authority and access to resources necessary to develop the plan that will allay their fears. They also have the credibility to interact effectively with industry leaders and other government jurisdictions.

Further motivation may come from legislative action. Several states have legislation similar to SARA Title III[2]. In several cases, it is more prescriptive[6,7]. Failure to comply may result in punitive action.

What are the responsibilities of local government leaders? Management of a crisis must be vested in an individual who is responsible and accountable. If this person does not have control and executive authority to direct the response, then he or she cannot be held either responsible or accountable. For practical purposes, alternates should be named for each defined position in the local emergency organization, as they are for the plant. Alternates must have the same responsibility and accountability as the individuals whose positions are assumed.

The question of direction and control must be confronted directly. It is impossible to establish a line of command and control while a crisis is in progress. The chain of command needs to be clearly laid down, understood, and universally accepted within an organization and among interfacing agencies and organizations long before an incident occurs.

The chief executive of a community is charged with coordinating the functions of the local fire and police departments (usually the first responders to an incident), the public works and public health departments, and any other agencies involved in aspects of emergency response. These groups may have differing views about their roles in managing an incident. It is up to the elected or appointed community leader to resolve these differences and ensure a coordinated approach. One caveat is appropriate here, as it was for the facility organization: The person filling the executive position may change, but the coordinating function may not. Continuity of leadership must be vested in the office, not the individual.

The Planning Team

Successful planning requires community involvement and support throughout the process. Those who actively participate in development of a program will accept it; a plan to which they have provided no input will probably fail. Cooperative interaction among responders and the community begins with the planning process.

Most important in the organization of the team that develops a community emergency preparedness plan is a leader who has the respect of the citizenry as well as of those community agencies and organizations directly involved in response to an emergency. The selected team leader should have both the time and resources available for performing the task from its inception through completion of the plan and thereafter. (It is a fact that a plan is out of date as soon as it is finished.) Management and communication skills are essential for gaining the cooperation of all concerned parties.

The team must have expertise in many areas or have ready access to it. Representatives of industrial facilities in the community that are potential sources of hazardous material releases should be included, as should knowledgeable officials from transportation systems: railroad, highway, marine, and air carriers. What matters most is that the group represent all elements of the community and be able to work cooperatively to reduce the risk associated with hazardous materials. Workable size is important.

If the number on the planning team is unwieldy, it can be pared by the creation of subcommittees that act in an advisory capacity.

The Planning Process

In the local sphere, as well as in the plant/facility, the planning process is the key to success. Agreeing to use a document prepared by others to follow in a "cookbook" fashion is to circumvent the planning process and diminishes the importance of those committed to it. The relationships of the industrial community, government, and the local community are often fragile. The process itself provides interaction among all participants so they become aware of their strengths and weaknesses, which are factored into the response mechanism to make it workable. The interaction has a positive effect of its own.

Planning Team Tasks

Hazards Analysis

Hazards identification, vulnerability analysis, and risk analysis together comprise the hazards analysis task that enables prioritization of the areas to be addressed by the community. However, it is important to note that help with this process is available. In 1987, the U.S. Environmental Protection Agency (EPA) developed a publication jointly with the Federal Emergency Management Agency (FEMA) and U.S. Department of Transportation (DOT) to assess the hazards related to potential airborne releases of extremely hazardous substances. Its title is *Technical Guidance for Hazards Analysis: Emergency Planning for Extremely Hazardous Substances*[27]. The initial task of a planning team is to establish priorities.

Hazards Identification

The first step should be to identify the hazards. This determines whether a plan is really needed, and for what. High priority hazards can be addressed before those with lesser impact potential. For facilities or transportation routes where the identified hazard is toxic or flammable material, the identity, location, and quantity must be precisely determined. The facility that manufactures, processes, stores, or uses such material is the logical source of this information.

Vulnerability Analysis

What threats do the identified hazards pose and to whom? The vulnerability analysis, sometimes called a consequence analysis, involves determination of the areas, populations, and facilities that may be at risk if a release occurs.

Risk Analysis

The team is now aware of the potential hazards and the area(s) of concern or vulnerability zone(s), but not the likelihood of any particular incident. This is determined in the risk analysis task, which ascertains the probability of accidental release and the severity of its consequences. Methodologies, both public and commercial, are available for obtaining estimates of the quantity of an EHS released (to air, especially), dispersion, and concentrations that could affect human health. The assumptions used in these as well as their capabilities and limitations should be studied carefully before any large investment is made.

The EPA *Technical Guidance* document[26] provides fairly detailed information useful for estimating the size of zones considered vulnerable because of the potential for acutely toxic effects of accidental EHS releases. Appendix D of that document gives additional information on suggested levels of concern and the bases for these. Local planners are encouraged to use one-tenth (0.1) of the National Institute for Occupational Safety and Health (NIOSH) Immediately Dangerous to Life and Health (IDLH) values as perimeter baseline levels[28]. When completed, a local vulnerability analysis should provide hard data usable for planning purposes:

1. Geographic description of the areas deemed vulnerable to the identified hazards, with the assumptions that led to the determination (e.g., release quantity, meteorological conditions).
2. The size and type of populations expected to be in the defined vulnerable zones, including special populations, with assumptions of time of day or night and season of the year.
3. Property, both public and private, and essential utilities and support systems that may be affected.
4. Environmental media that may be affected, and how.

Examples of emergency planning information that result from the hazards analysis process include needs for facilities and equipment; facilities from which to conduct response operations and to which people can be temporarily relocated if necessary; equipment for accident mitigation, emergency worker protection, and spill cleanup. Also, criteria for determining the extent of emergency response required can be established (i.e., the levels of severity of various emergencies that would require particular consequent notification, communication, and protective actions). For example, what constitutes a limited emergency condition for which a lower level of response activity is needed, and what is a full emergency condition demanding full resource activation. Emergency action levels (EALs) or an incident classification system are preestablished conditions that can be used to trigger a desired response. Some industries and jurisdictions use the classification system: alert, site emergency, general emergency. Whatever the name, EAL definition removes the ambiguity of a wait-and-see attitude when a problem is suspected.

Resources

Resources, in terms of people as well as facilities and equipment, are necessary for the contingency plan to work. As above, the development and use of questionnaires to construct matrices of available resources is useful. The questionnaires should be provided to the sources of identified hazards (facilities, transporters) and to local response and government agencies. Care should be taken not to overburden personnel of these organizations with questionnaires and requests for detailed listings. Be sure that they understand the need for the requested information. The National Response Team's Hazardous Materials Emergency Planning Guide (NRT-1)[29] contains a list of questions in its section titled "Capability Assessment."

Personnel

The people available to implement the contingency plan must be identified before an emergency responsibility matrix can be constructed. Response to the preliminary questionnaire provides initial indication, and further detail is acquired in individual agency or organization meetings with the planning team. At these meetings, the specific community points of contact should be identified by position title, with their areas of responsibility. A list of the individuals who hold these positions, and their alternates, should be developed separately, with the chain of command shown on an organization chart. As noted earlier, positions stay fairly constant, but the people who fill them often do not. Therefore, the plan should contain position titles only, with names and 24-hour phone numbers in a separate (yet readily accessible), easily updated document or appendix. Once the personnel resources and areas of responsibility are identified, a matrix of groups versus functions is readily constructed.

Facilities

In most cases, local governments already have facilities in place to handle the types of emergency situations they are likely to face, such as severe weather-related problems. Ideally, procedures are in place for their activation and use. It should be a relatively simple matter to augment these existing resources as necessary for industrial emergency response. Pre-planning is necessary, however. Letters of agreement or memoranda of understanding must be executed between government leaders and the agencies, organizations, or individuals responsible for buildings or portions thereof that may be needed during an emergency.

Facilities need not be dedicated for emergency use. In many communities, a section of the town hall or police or fire department headquarters is set aside to store the equipment necessary for rapid setup of an efficient center from which to direct emergency response activities. The equipment needed in this Emergency Operations Center, or EOC, is described in the next section.

The public receives most of its information about an emergency situation through the media, primarily the electronic media. For this reason, an area separated from (but in contact with) the EOC should be designated for their use. It is best if this media center is staffed by spokespersons from the accident site as well as from local government and response agencies. It is best to ensure that sufficient communications equipment is provided and that the needs of print media representatives are accommodated.

Other facilities that may be needed in local response to an emergency depend on what has been identified during the hazards analysis phase. If emergency response personnel may be exposed to toxic or radioactive materials, a decontamination center may be required. This can be simple or elaborate, depending on the need. A garden hose and wading pool–type container for runoff water (which cannot be allowed to enter groundwater or sewage systems untreated) may be sufficient. Portable, inflatable tents have been designed for this purpose, as have mobile vans. If wastewater can be contained or treated before disposal, a shower in a fire station may be adequate. Whatever means are employed, procedures should be in place for their use. Contaminated injured persons should be decontaminated before receiving medical care, if possible, to avoid cross-contamination of emergency medical personnel. Consideration should also be given to the possible need for vehicle decontamination. Some commercial car washes recycle their wastewater, and agreements could be made with them.

Emergency medical treatment must be available. A triage area should be set up near the scene of an accident for first-aid treatment of the injured by specially trained nurses or emergency medical technicians. Ambulance companies with which agreements have been prearranged are directed to this area to transport the severely injured to area hospitals or treatment centers, with which prior agreements have also been made. The planning team should ensure that medical personnel at the designated centers are aware of the potential health hazards in the community before an incident occurs.

The public protective action of choice in typically fast-moving industrial accidents is sheltering in place, that is, remaining or going indoors and shutting off outside ventilation. However, evacuation of areas near the scene may be necessary. Besides an evacuation plan that describes optimal routes, reception or relocation centers should be identified for evacuees. Experience has shown that up to 80 percent of these will travel to the homes of friends or relatives out of the area. Therefore, reception or relocation centers are planned to handle no more than 50 percent of potential arrivals. Public schools are often designated as centers, because they have cafeterias, adequate sanitary facilities, and large open gymnasiums. Evacuees are temporarily housed in a relocation center. Sleeping accommodations (cots, blankets, etc.) are usually provided by volunteer organizations such as the Red Cross. A reception center may serve the same purpose in a small or short-term evacuation. In a large-scale evacuation, the reception center typically serves as a registration center and, if necessary, a decontamination center. From that location, people go to predesignated relocation centers for longer term care. The reception center in this case documents the location of evacuees and functions as an information center for concerned relatives.

Equipment

The equipment needed for emergency operations at the local level is to some degree generic, yet also hazard-specific. Emergency Operations Centers (EOCs) are equipped to handle any kind of major emergency. Communications equipment will be essentially the same in all cases, as will public warning systems and notification methods, traffic and access control, public works, law enforcement, and health and medical services. Copy and telefax equipment should be available for hard-copy transmission and reception of data and messages.

Large-scale maps of the planning area should be prominently displayed in the local EOC. Major transportation and evacuation routes as well as identified hazard locations with their vulnerable zones are examples of the information that should be provided on the base map or on overlays to it. Circles representing radial distances from a central point may be added for major hazards or, preferably, used as separate overlays to focus on specific areas. The latter method is especially useful where several hazard sites have been identified, to avoid cluttering up the base map and eliminating the need for revision when conditions change. Another helpful overlay shows the location of special facilities and impaired individuals who may need assistance in an emergency. Airborne dispersion plume projection overlays or templates are useful additions, especially for transportation accidents; with known wind speed and direction, populations at risk are identified promptly for protective action.

Specialized equipment for response to industrial plant emergencies depends on the nature of the identified hazards. Much of this necessary planning information comes from the hazards analysis process.

All of the equipment need not be owned by the local jurisdiction. Other jurisdictions, local industries, and cleanup contractors may be able to assist. For liability considerations, it is important that letters of agreement or memoranda of understanding be executed in these cases. They should clearly spell out the limitations of the agreement as well as capabilities, and provide 24-hour contact names and phone numbers.

Content of the Plan and Procedures

The best local level contingency plan encompasses all known natural and technological hazards and is adaptable enough to accommodate those identified in the future. To call it an "all hazards" plan may be overly optimistic. Thus, the term multihazard emergency operations plan (EOP) is recommended. Such a plan comprises a basic plan providing an overview of the jurisdiction's approach to emergency management, functional annexes in support of the basic plan to address specific activities critical to emergency response and recovery, and hazard-specific appendices to the plan that contain technical information, details, and procedures or methods for use in specified emergency situations. Examples of these appendices are SARA Title III hazardous materials transportation accidents or incidents at subject facilities, severe weather emergencies experienced in the area, and nuclear power plant emergencies. FEMA's planning guide[26] also recommends an appendix addressing nuclear attack. Dealing with the aspects common to all hazards first and then examining hazard-unique characteristics is both efficient and economical.

The Basic Plan

If the local jurisdiction chooses not to develop a multihazard EOP and instead prefers a single-hazard hazardous materials contingency plan, the elements described below in the basic plan and functional annexes must still be in place.

Basic Plan

The basic EOP is an umbrella plan that contains a substantial amount of the generally applicable organizational and operational detail. It establishes the structure reflected in annexes. The basic plan cites the legal authority for the plan, summarizes the situations addressed, explains the concept of operations, and describes the organization and responsibilities for emergency planning and operations.

The basic plan should also include maps, organization charts, and the emergency responsibility matrix. Format is not as important as ensuring that all known contingencies are addressed as simply as possible. However, a format compatible with neighboring jurisdictions is recommended.

Functional Annexes

The generic functional annexes define and describe the policies, procedures, roles, and responsibilities that are inherent in the functions before, during, and after an emergency. They should be sufficient to cope with any unforeseen emergency.

The functional annexes serve as the basis for, and may well include, standard operating procedures (SOPs). These are user-friendly, checklist-type instructions for the various segments of the emergency response organization to execute the functions defined in the annexes. A telephone roster listing the names and phone numbers (home and business) of key members of the emergency response organization (and their alternates) should be provided in controlled copies of the EOP, in a separate section for easy update and revision.

One area too often overlooked in the local planning process, as it is for facilities, is the recovery phase, that is, the steps taken to return to normal following an emergency. The planning team would be wise to visit a community where an accident has previously occurred to learn from them what recovery problems they faced and how they resolved them.

Hazard-Specific Appendices

The unique characteristics of hazards identified for the jurisdiction are included as appendices to the functional plan. An appendix for a particular hazard may include one, some, or all of the functional annexes, depending on recognized need. A single appendix may address all functional annex considerations related to a particular hazard.

Plan Integration

Coordination of contingency planning between industry and community, and among authorities of neighboring jurisdictions, is necessary to develop mutually acceptable solutions to perceived problems. Should an actual emergency response be necessary, cooperation and commitment supply the means for orderly, timely decision making.

It takes time to lay the groundwork among the members of these partnerships—local government, community, and industry—to establish an innovative approach to cooperative problem solving. Effective listening, the open exchange of ideas, understanding of the roles of participants from these diverse sectors, and strong leadership are the critical ingredients for success. Industry should provide personnel to local planning teams as required by Title III, and community planners should be invited to industry planning meetings.

PUBLIC INFORMATION

Public information has two faces in contingency planning: education about the plan itself and why it was developed, and notification of an emergency condition. The first is a public relations function; the second a necessary part of the plan itself, contained in the applicable functional annexes.

Public Education

Residents and businesses in industrial areas are increasingly aware of potential threats to their well-being from industrial and transportation accidents. They are more concerned about major disasters than they are about the host of minor incidents that occur. Community and employee right-to-know legislation is helping to ensure this but should not be construed as negative. The more information citizens have about environmental conditions in their communities, the better equipped they are to participate in measures for their own protection from unacceptable safety and health risks. The hazards in a community and what both industry and the jurisdiction are doing to minimize the risks and manage emergency situations that may arise must be made known to them clearly and explicitly.

Perception and truth can be the same in the public eye, however subjective, inconsistent, and irrational this may be. People react differently to the same risk, depending on their backgrounds and their level of risk acceptance. Voluntary risks such as smoking and not wearing a seat belt are usually accepted, whereas the involuntary risks of exposure to asbestos, contaminated drinking water, or a toxic plume are not. Health risks, especially long-term, are of primary concern to those who resent risks not of their own choosing. While risk comparisons may be valid, it is better to focus discussion on preventive measures, emergency preparedness, and containment and remediation procedures.

The public gets most of its information through the media, which can sometimes oversimplify complex situations. The key is to present essential factual information in readily understandable terms—that is, without technical jargon. Also, building rapport ahead of time can lead to fair treatment in the event of an accident. Press releases and conferences during the planning process help to accomplish this goal.

When the first round of planning is complete and the plan itself approved, a familiarization program should be undertaken so that citizens will understand their expected actions in the identified potential emergency situations. Presentations to community groups are good, but they may not reach all who could be affected. Explanatory brochures are often used, but these are apt to be misplaced or discarded. Experience has shown that readily accessible emergency information presented positively and in an attractive format is remembered and used. One recommended method is the creation of an attractive calendar distributed annually to households and commercial establishments that contains simple instructions for citizens to follow, with graphical representation of preferred evacuation routes and explanation of warning signals. Some industries have provided the information on one- or two-page inserts in local telephone directories. In agricultural areas, special information on the care of crops and animals should be given. Public confidence is enhanced when citizens have the factual information needed to make intelligent decisions.

Emergency Public Information

When an emergency does occur, the local emergency response team must be promptly notified, and a public warning issued to all who may be affected. Initial notification of a problem to the emergency response team may be made by beeper, radio, or telephone. The Initial Notification and Public Alert and Warning functional annexes to the basic emergency operations plan should contain the methods and procedures for these processes. A standardized notification message form should be available to both sender and receiver of the initial information.

How the media are treated while an emergency is in progress determines, to a large extent, public perception and reaction. Establishment of a media briefing center or public information center was described earlier. Here the local designated spokesperson can coordinate the timely provision of accurate, detailed, and meaningful information to media representatives who, because they are familiar with the contingency plan through advance preparation, will present the situation more fairly than if they had no prior knowledge.

SELECTED REFERENCES

1. 29 CFR 1910.1200—Occupational Safety and Health Standards, Section 1910.1200, Hazard Communication, 52 FR 38152, August 24, 1987 as amended.
2. 42 USC Sections 110011-11050. Emergency Planning and Community Right-to-Know Act of 1986, Public Law 99-499, October 17, 1986 as amended.
3. 29 CFR Part 1910.120—Occupational Safety and Health Standards, Section 1910.120, Hazardous Waste Operations and Emergency Response, 54 FR 9294, March 6, 1989.
4. 29 CFR Part 1910.138—Occupational Safety and Health Standards, Section 1910.138(a), Emergency Action Plan.
5. 29 CFR 1910.165, Occupational Safety and Health Standards, Section 1910.165, Employee Alarm Systems.
6. New Jersey Register, June 20, 1988, Chapter 31, *Toxic Catastrophe Prevention Act Program,* Cite 20 N.J.R. 1402.
7. California Health and Safety Code, Chapter 6.95, Section 25500 et seq. 1985 Cal. Stat. 1167.
8. Community Awareness & Emergency Response (CAER), 1986, *Site Emergency Response Planning,* Chemical Manufacturer Association, Washington, D.C.
9. Burns, C. C., personal communication with confidential client, 1990.
10. Armenante, P. M., 1991, *"Contingency Planning for Industrial Emergencies",* Van Nostrand Reinhold, New York.
11. Burns, C. C., personal communication with person who viewed the telecast, 1988.
12. Burns, C. C., personal communication with confidential clients, 1990.
13. Burns, C. C., personal communication with confidential client, 1987.
14. Lees, F. P., 1980, *Loss Prevention in the Chemical Industries,* Volumes 1 and 2, Butterworths, London.
15. *NFPA 19, Specification for Motor Fire Apparatus,* 1984, National Fire Protection Association, Quincy, MA.
16. *NFPA 193, Fire Department Ladders—Ground and Aerial,* 1984, National Fire Protection Association, Quincy, MA.
17. *NFPA 17, Dry Chemical Extinguishing Systems,* 1985, National Fire Protection Association, Quincy, MA.
18. Schultz, N., 1985, *Fire and Flammability Handbook,* Van Nostrand Reinhold, New York.
19. *NFPA 11A, High Expansion Foam Systems,* 1983, National Fire Protection Association, Quincy, MA.
20. Prugh, R. W. and Johnson, R. W., 1988, *Guidelines for Vapor Release Mitiga-*

tion, Center for Chemical Process Safety. American Institute of Chemical Engineers, New York.

21. Froebe, L. R., 1982, United States Local Government Plans, in *Hazardous Materials Spill Handbook* (G. F. Bennett, F. S. Feates and I. Wilder, Editors), McGraw-Hill, New York.

22. Chlorine Institute, 1986, *The Chlorine Manual,* 5th edition, The Chlorine Institute, Washington, D.C.

23. Katz, W. B., 1982, Plant Operations, in *Hazardous Materials Spill Handbook* (G. F. Bennett, F. S. Feates and I. Wilder, Editors), McGraw-Hill, New York.

24. Rome, D., 1982, Personnel Safety Equipment, in *Hazardous Materials Spill Handbook* (G. F. Bennett, F. S. Feates and I. Wilder, Editors), p. 13-1 to 13-21, McGraw-Hill, New York.

25. Hanna, S. R. and Drivas, P. J., 1987, *Guidelines for Vapor Cloud Dispersion Models,* Center for Chemical Process Safety. American Institute of Chemical Engineers, New York.

26. CPG 1-8, Federal Emergency Management Agency, *Guide for the Development of State and Local Emergency Operations Plans,* October 1985.

27. U.S. Environmental Protection Agency, Federal Emergency Management Agency, U.S. Department of Transportation, *Technical Guidance for Hazards Analysis,* U.S. Government Printing Office: 1988 519-501/63067.

28. idem, Appendix D

29. NRT-1, National Response Team, *Hazardous Material Emergency Planning Guide,* Washington, D.C., March 1987.

SUGGESTED READING

1. Battling crimes against nature. *Time* 135(11). March 12, 1990.

2. Junkyard owner fined $50G. *The Patriot Ledger,* Quincy MA. March 9, 1990.

3. *Community Right-to-Know Manual,* Thompson Publishing Group, January 1990.

4. 29CFR Part 1910—Occupational Safety and Health Standards, Section 1910.120, Hazardous Waste Operations and Emergency Response, Federal Register, 54(42), March 6, 1989.

5. *Guide for the Development of State and Local Emergency Operation Plans* (CPG 1-8), Federal Emergency Management Agency, October 1985.

6. U.S. Department of Transportation, *Hazardous Materials Emergency Response Guidebook* (DOT-P-5800.5), 1990.

7. U.S. Environmental Protection Agency, 40CFR Part 300—National Oil and Hazardous Substances Pollution Contingency Plan, July 1988.

8. Awareness and Preparedness for Emergencies at the Local Level (APELL), United Nations Environment Programme, Paris 1988

9. American Chemical Society, Chemical Risk Communication, 1988.

10. Chemical Manufacturer's Association, *Community Awareness and Emergency Response Program Handbook,* 1986.

17

Risk Financing

Michael J. Natale

Not even the most sophisticated risk management program can completely eliminate an organization's potential for financial loss from unpredictable circumstances. While risk control measures can reduce the frequency or severity of losses, risk financing tools must be used to fund losses that do occur. Insurance is the most familiar means of risk transfer, but it is only one of many techniques the risk manager uses to fund losses. Others include

- Self-insurance
- Cash flow plans
- Fronting plans
- Banking plans
- Captives
- Risk retention groups and risk purchasing groups.

Other than pure "first-dollar" (i.e., no deductible) insurance coverage or pure self-insurance, each technique generally involves a combination of insurance and self-insurance.

INSURANCE VS. SELF-INSURANCE

"If we had a first-dollar insurance policy, the insurance company would have paid for that loss. Now, it comes out of current operating funds." Although this statement is true, a quick examination of the way insurance companies determine premium rates will show why insurance coverage without deductibles is not necessarily cost effective in the long run.

An insurance premium has three basic components: (1) funds set aside to indemnify the insureds for losses; (2) funds for operating expenses; and (3) funds for profit and contingencies. When determining the specific premium, the insurer will request a record of losses over the past few years. The insurer then calculates what the expected losses will be based on prior loss experience. To that figure are added expenses and profits to arrive at the total premium. When an organization self-insures, it eliminates the profit component, acquisition expense, and underwriting expense. Of course, losses and loss adjustment expenses must still be paid.

Loss adjustments, which include evaluation, negotiation, and settlement of claims, can be made by various types of outside claims administrators. If claims activity warrants it, claims specialists can be added to corporate staff, and the company can self-administer, as well as self-insure.

The extent to which an organization insures or self-insures is influenced by insurance market conditions and the financial strength of the organization itself. Insurance availability has proven to be cyclical, with multiyear periods during which insurance markets have been "hard" (limited, high-priced coverage) and "soft" (easily obtainable, low-priced coverage). During hard markets, many organizations seek creative alternative risk financing techniques, and even with the return to soft market conditions, they continue to maintain and refine self-insurance techniques in lieu of insurance.

As with any major purchase, the benefits of insurance must be measured against its costs. Benefits include

- Indemnification for losses
- Reduction of uncertainty
- Insurer-supplied loss control services.

When the costs of providing these benefits become excessive, organizations seek alternative cost-effective risk financing methods.

In the chemical process industry, various forms of self-insurance became attractive because of the "long-tail" nature of certain liability exposures. Product liability, for example, generally requires many years before it manifests itself, and any resulting claims can take many additional years to wend their way through the settlement and court adjudication process. Since insurers can be liable for a claim many years after the expiration of their policy, they must collect premiums far in advance of potential payouts. For example, under an occurence form of liability policy, if a drug is manufactured in year X when the policy is in effect, but a claim resulting from that drug is not lodged until five years later, the policy in effect during year X must still respond. Although there may be other contributing policies, if the drug that caused the injury was manufactured during year X, the policy in effect during year X will be one of the contributors to the claim payout. Insurers must take this into account and charge some estimated premium, discounted for the time value of money, for future claims payments. Under a self-insurance plan, the company can retain the funds until the claim must be paid and can use the funds in the interim to generate increased profits. This is the major benefit of self-insurance.

EXPOSURES

Before considering alternatives to pure insurance or self-insurance, the plant manager should be familiar with the exposures inherent in the chemical process industry. These exposures, separately or in combination, are candidates for insurance, self-insurance, or alternative risk financing techniques. Many of the following exposures are common to all industry, but those that may cause additional concern for the chemical process industry are covered in more detail. The following exposures will be discussed in the order of the degree of hazard they represent to the chemical process industry:

- Product liability
- Environmental impairment liability
- Property
- Boiler and machinery
- Workers' compensation
- Business interruption
- Directors' and officers' liability
- Vehicle liability
- General liability
- Inland marine
- Fidelity.

Product Liability

Any product has the potential to be manufactured or designed in a negligent manner. In some cases, courts have held manufacturers responsible for injuries arising out of defective products without regard to negligence. This finding of strict liability creates serious problems for defendants, because the defense of proper and reasonable care cannot be used successfully.

Pharmaceutical manufacturers have been particularly affected by product liability cases in recent years. Product tampering has been one cause for concern. While the loss control method of tamper-resistant packaging has proven effective, it is still prudent to amend general liability policies to address the cost of product recall and the subsequent loss of income to the company. Also, previously undiscovered side effects of certain drugs can lead to many claims being filed far into the future.

Paint manufacturers that develop and distribute specialized paints for industrial application and produce coatings such as fire-resistive paints intended to withstand environmental conditions may be subject to product liability actions if the paints do not perform according to specifications. Although the Poison Prevention Packaging Act and the Hazardous Substances Act require child-resistant packaging and warning labels, poisonous synthetic-based lacquers and varnishes still pose poisoning hazards to children.

Environmental Impairment Liability

The chemical process industry is subject to environmental impairment liability in several different areas. The disposal of toxic substances such as benzene, cyanide, lead, phenol, and chloroform can subject the chemical processor to liability. Company-operated treatment facilities can remove most organic materials through the use of solvent recovery techniques and effective wastewater treatment. Improper disposal or unanticipated leaks or discharges can, however, result in liability for bodily injury, property damage, and cleanup costs.

Even after toxic chemicals leave the site, the company may still be responsible if the disposal contractor does not dispose of the waste in a prescribed manner. The federal Superfund Act is very broad and can ascribe liability and responsibility for cleanups to a wide variety of responsible parties.

Other environmental exposures include sensory pollution via fumes and odors and potential spills from railroad cars or trucks that service the chemical processing facility.

In one of the most dramatic environmental impairment cases on record, a gas leak at the Union Carbide plant in Bhopal in 1984 killed 3,500 people and injured thousands more. The suit that was brought on behalf of the plaintiffs sought $3 billion.

Insurance coverage for environmental impairment liability became increasingly difficult to obtain in the 1980s. Creative use of alternative financing mechanisms such as captives and risk retention groups should play an important role in the 1990s.

Property

The use of solvents, compressed gases, and sophisticated machinery and pressure vessels can subject buildings and contents to catastrophic loss. Industrial-scale mixing and milling operations, compressed gases, or any process under pressure can result in static electricity buildup and explosion. The fire hazard becomes more critical when large amounts of flammable substances are stored and processed.

Boiler and Machinery

Use of sophisticated equipment and machinery requires analysis of probable maximum losses resulting from an accident involving these items. Traditional property insurance policies do not cover these circumstances, and separate, specialized policies are usually written. The boiler and machinery hazard is not limited to steam boilers but also includes similar equipment subject to internal pressure, such as air tanks, compressors, furnaces, and refrigeration systems. Damage caused by accidents involving flywheels, electrical machinery, and turbines are included in this category of risk.

Due to the potential for catastrophic loss involving pressurized equipment, extensive loss control efforts are usually undertaken.

Workers' Compensation

In workers' compensation, negligence and fault on the part of the worker or employer are not considered when determining the benefits to be paid. All work-related accidents or injuries are compensated under the law. Workers' compensation is thus a no-fault system for injured workers. Payments are made for medical expense, lost income, rehabilitation benefits, and death benefits. Workers are subjected to numerous chemicals that may be biologically active. If these chemicals are absorbed through the skin, inhaled, rubbed into cuts, or swallowed, they can cause injury or disease. Workers injured in this manner are eligible to file workers' compensation claims. This exposure is in addition to the sprains, cuts, back injuries, and other accidents inherent in other manufacturing industries, and that are also present in the chemical process industry.

Business Interruption

A business interruption loss is usually concurrent with a major property loss. Business interruption is generally covered on a standardized form known as the Business Income Coverage Form. For coverage purposes, business income is defined as (1) net income (net profit or loss before taxes) that would have been earned or incurred; and (2) continuing normal operating expenses, inlcuding payroll.

Factors to be considered in determining how much coverage is necessary are (1) How long will it take to rebuild the facility? (2) Is there another plant capable of taking on increased capacity? (3) Can some of the processing be contracted out? Extra expense coverage should be considered here to provide funds to allow the firm to continue its processing activities temporarily in some other way. These expenses might include rental of temporary premises or equipment to continue operations.

Directors and Officers (D & O) Liability

Directors and officers are expected to exercise prudence and good faith in carrying out their managerial responsibility. A breach of this duty can lead to a D & O claim. Lack of good judgment, for example, in allowing a pollution episode to develop can lead to an action. There are three sources of litigation: (1) derivative action suits brought by shareholders on behalf of the corporation, (2) representative action suits brought by shareholders against the corporation on their own behalf, and (3) third-party actions brought by outside interests such as creditors.

Vehicle Liability

Use of passenger cars, trucks, and vans presents exposures not atypical to all industrial organizations. Funding must be available if the use of these vehicles results in an accident that causes bodily injury or property damage to a member of the public. Funds must also be available to repair the damaged company vehicle. Any employees injured as a result of driving a company vehicle are covered under workers' compensation. The unique concern of the chemical process industry is the possible use of tractor-trailers to haul potentially polluting substances where a single accident can result in a vehicle liability exposure, workers' compensation, and an environmental impairment liability exposure, which could cause bodily injury or property damage and result in cleanup costs.

General Liability

General liability involves responsibility for bodily injury or property damage to members of the public arising on the company's premises or arising out of company operations. A trip-and-fall on the premises would be considered a general liability accident. Certain types of liability, such as product liability, environmental impairment liability, and vehicle liability are usually treated as separate exposures from the catchall category of general liability.

In most chemical processing facilities, the public does not have ready access to the premises. The presence of contractors, salespeople, and others who visit can create general liability exposure. Wastewater treatment or pretreatment facilities and large electric transformers can create an attractive nuisance hazard that should be monitored and carefully secured.

Inland Marine

This esoteric-sounding exposure is generally related to transit coverage, although coverage for data processing equipment and valuable papers and records are included in this unusually titled category. With regard to transit, even if common carriers are used, separate funding should be considered, because a common carrier is not responsible for perils such as an act of God, an act of war, exercise of public authority, or inherent defects in the goods being shipped.

Fidelity

Fidelity refers to any fraudulent or dishonest act of an employee, inlcuding such acts as larceny, embezzlement, theft, and forgery. Although this exposure rounds out a discussion of the hazards facing the chemical process industry, the industry does not possess any extraordinary fidelity exposure. It is described so as to not be overlooked when consideration is given to overall program financing.

ALTERNATIVE RISK FINANCING TECHNIQUES

Having gained an understanding of the general concepts of insurance and self-insurance as well as surveying the general hazard categories to which the chemical process industy is subjected, we now turn our attention toward alternative risk financing techniques. Many of these techniques provide a combination of insurance and self-insurance components.

Cash-Flow Plans

1. Paid Loss "Retro"—Under a retrospectively rated plan, the ultimate premium is dependent on losses. Retrospectively rated plans base the insured's premium on actual losses plus the cost of related insurer expenses and profit. These plans differ from conventional (guaranteed-cost) insurance in the manner of premium payment. Under conventional insurance, the insured pays an advance premium based in part on estimated ultimate losses. Under a paid-loss plan, the insurer accepts a reduced initial premium pay-in, with the balance secured by some type of financial security such as a letter of credit.

The insurer pays claims from the initial deposit as they arise, and the funds are replenished by the insured at required intervals. Cash-flow benefits accrue to the insured, thus allowing some of the benefits of self-insurance.

2. Incurred Loss Retro—An incurred loss retro operates like a paid loss retro, but the cash-flow advantages are not as favorable. Under this type of plan, the premium is also dependent on losses. Rather than being based on paid losses, however, the premium is based on incurred losses, which include not only paid amounts but also estimated outstanding amounts. Because an estimate for the outstanding amounts is included in the premium, the cash-flow benefits to the insured are not as great as in a paid loss retro.

3. Split Payment Plans/Insured Cash Flow Plans/Premium Payment Plans—These plans do not have a fixed definition, except they all provide some type of cash-flow benefit to the insured. Under each type, there is an initial payment, with adjustments at the end of the policy period. The structure and cash-flow arrangements are limited only by the creativity of the insureds and their tax advisors.

4. Compensating Balance Plans—Under this plan, the insured agrees with its bank and the insurance carrier to purchase an insurance policy and pays the insurer the entire estimated standard premium at policy inception. The insurer is entitled to its share of the premium; it witholds non-loss-premium, and deposits the remainder in a non-interest-bearing account in its own name at the bank where the insured maintains a line of credit. The insurer's account qualifies as a portion of the insured's compensating balance, which must be maintained for the bank to grant the insured its lines of credit, and contains funds segregated from the other deposits of the insured. Upon the insurer's deposit of the net premium, the insured is permitted to withdraw a corresponding amount from its required compensating balance. The insurance carrier withdraws funds from the bank account it established to pay developing losses. As in a paid-loss plan, the insured is billed periodically for reimbursement of paid losses. When billed, the insured reimburses the insurer by depositing the billed amount into the insurer's account.

This type of arrangement is designed for insureds that must keep substantial compensating balances to support lines of credit.

Fronting Plans

In some cases, state law makes self-funding difficult or impossible for certain lines of coverage. In these cases, an insurer can be paid a fee to issue required certificates or other documents. The insured then signs a letter of indemnity wherein it agrees to pay the insurer for any losses incurred. While this arrangement may sound as if it is circumventing the law, it has been found acceptable by regulators, because the financial stability of the insurer backs up the insured. The cash-flow benefits of self-insurance still rest with the insured.

Banking Plans

Banking plans have been designed to spread losses over long periods of time. They provide the cash-flow advantages of self-insurance and can build a reserve fund for losses. There is an initial negotiated premium, with the final cost being determined by a retrospective formula that can be as long as 10 years, with annual adjustments. The major benefits of this type of plan are the stabilization of earnings by spreading what are essentially self-insured losses over a long term, and the possibility of tax deductibility that might not be obtainable under other forms of self-insurance.

Captives

A captive is a subsidiary owned by one or more organizations. Its primary purpose is insuring the exposures of its owners or its owners' affiliates. Captives have been developed so that coverages not readily available in the traditional marketplace could be obtained at a reasonable cost. Owners have been able to improve cash flow and earn investment income that normally accrues to insurers. Bermuda has long been a popular domicile for captives because of its regulatory atmosphere. Capitalization, reporting and auditing requirements, and investment restrictions are structured to encourage captive formation. In the 1980s, domestic captives became more popular in states such as Vermont and Colorado, where regulatory burdens were relaxed. Tax laws with regard to captives are extremely complex. IRS rulings and court decisions concerning tax deductibility questions continue to evolve rapidly.

Risk Retention Groups and Risk Purchasing Groups

Because individual states can regulate insurance due to the McCarran-Ferguson Act, federal legislation was desirable for the formation of these groups, which crossed state boundaries. The federal Risk Retention Act authorized their formation and provided for less burdensome licensing requirements.

A risk retention group must be licensed as an insurance company under the laws of one state. It can provide insurance only to members, each of whom has a partial ownership interest. The owners and members must be involved in similar businesses. Once a group is licensed in one state, it can write insurance coverage for members from other states without being licensed in those states. Many risk retention groups are actually group captive insurance companies. The relaxed licensing requirements have made this an attractive risk financing alternative.

The purpose of a risk purchasing group is the purchase of liability insurance on a group basis. Prior to passage of the McCarran-Ferguson Act, many states forbade the offering of liability insurance on a group basis. The regulation of a purchasing group and its insurer is subject to the laws of the state where the group maintains its principal place of business. A purchasing group purchases coverage for its members, while a risk retention group spreads the liability exposures of its members.

SUMMARY

The periodic unavailability of certain insurance coverages coupled with high prices of others has led many companies to explore the alternative risk financing techniques described in this chapter. The selection of an alternative generally requires an in-depth feasibility study. While many specious rules of thumb have been promulgated, insurance market conditions in the long run, corporate loss history, corporate financial size and condition, and management's philosophy toward the assumption of risk are the crucial factors in selecting an appropriate self-insured retention. The approach should not be dictated by artificial formulas or what other companies in the same industry do, but rather guided by what other companies with similar financial profiles and philosophies have found to be effective.

Once an organization has experienced the cash flow benefits of some form of self-insurance, it will often retain a program that is self-funded to some degree even if the insurance markets soften. The plant manager must maintain a constant awareness of the insurance marketplace and newly develping alternatives in order to ensure that the risk financing option selected remains the best available.

SUGGESTED READINGS

1. Tiller, M. W., Blinn, J. D., and Kelly, J. J. 1988. *Essentials of Risk Financing,* Vols. I & II. First ed. Malvern, PA: Insurance Institute of America.
2. Warren, D., and McIntosh, R. 1991. *Practical Risk Management—The Professonal's Handbook*. Oakland, CA.:Warren, McVeigh & Griffin.
3. The National Underwriter Company. 1991. *FC&S Bulletins*. Cincinnati, OH: The National Underwriter Company.

4. International Risk Management Institute, Inc. 1991. *Commercial Liability Insurance,* Vols. I & II. Dallas, TX: International Risk Management Institute.
5. International Risk Management Institute, Inc. 1991. *Risk Financing—A Guide to Insurance Cash Flow.* Vols. I & II. Dallas, TX: International Risk Management Institute, Inc.
6. Lilly, C. C., III, and Boggs, H. G., II. 1991. *The Self Insurance Manual,* Vols. I & II. Chatsworth, CA: NILS Publishing Company.
7. Buraff Publications. 1989. *Insurance and Risk Management for Business and Government.* Washington, DC: Buraff Publications.

18

Computer Techniques

Kenneth F. Reinschmidt

Among the most significant computer technologies that have a major impact on chemical and process plant risk assessment are expert systems. Combined with computerized plant data bases, expert systems are assisting plant operational and safety personnel with situation diagnosis, spill evaluation, crisis management, risk assessment, and other related functions.

Expert systems are tools for improving productivity through knowledge. As an indication of the immediate practical value of this emerging computer technology, a survey of more than 100 expert systems applications in the manufacturing industry reported economic gains of $10,000 to $2,000,000 per year. The nontangible benefits of these expert systems include wider distribution of expertise within the organization, improved decision-making, and enhanced performance of personnel and equipment[1].

As one example of expert systems implementation, the Dupont Corporation reportedly has 200 expert systems, virtually all on personal computers, with hundreds more under development. The average payback is reported to be $100,000 per year per expert system, for an average of one man-month of development cost. Dupont reports that the average annual return on investment from expert systems is 700 percent[2].

Risk assessment and crisis management are fertile fields for the application of expert systems, because experts are not always available at the right place at the right time. Many activities are performed at locations far from the source of expertise. Maintaining full-time experts at each plant site is often not cost effective, and decisions must often be made quickly, without time to bring in outside consultants. Regulations may play an important role in the assessment of risk and the response to crises, and people on the spot may be unfamiliar with federal, state, and local codes and regulations.

Risk assessment expertise is often heuristic, that is, based on experience and judgment rather than on theory and analysis, and is, therefore, especially suitable for representation by expert systems. For these reasons, chemical and process firms should consider expert systems to provide computerized access to expert knowledge.

DEFINING EXPERT SYSTEMS

Expert systems are part of the field of artificial intelligence. Artificial intelligence is notoriously difficult to define, so this definition from *The Handbook of Artificial Intelligence*[3] will suffice for our purposes:

> Artificial intelligence (AI) is the part of computer science concerned with designing . . . systems that exhibit the characteristics we associate with intelligence in human behavior—understanding language, learning, reasoning, solving problems, and so on.

Expert systems are not concerned with understanding language or other aspects of intelligence, but are concerned only with solving problems. Unlike other methods of artificial intelligence, expert systems are not based on any general theory of intelligence or universal method of problem solving, but rather are based on the straightforward assumption that intelligence lies in knowledge. That is, people who know more exhibit more intelligence (can solve more problems) than those who know less. Consequently, it follows from this premise that, if enough knowledge can be represented in a computer form, the computer will display intelligence-like behavior.

The first expert system was the program MYCIN, developed in the early 1970s at Stanford University to assist medical doctors in the diagnosis of bacterial infections in their patients. From this beginning, many diagnostic expert systems have been developed, as the reasoning involved in diagnosis of the causes of problems is suitable to representation by expert systems[4].

Whether an expert system in fact replicates the reasoning process used by a human expert is primarily of interest to computer scientists; as a practical matter, it is only necessary that the expert system replicate the results that the human expert would have obtained from the same information. An expert system can perform no better than the experts who created it; human experts are fallible; therefore, expert systems are fallible.

Expert systems, then, are intended to replicate the problem-solving capabilities of human experts. Or, to quote a definition given by the National Academy of Sciences:

> Expert systems consist of a body of knowledge and a mechanism for interpreting this knowledge. The body of knowledge is divided into facts about the problem

and heuristics or rules that control the use of knowledge to solve problems in a particular domain.

In the last few years, expert systems have become the most visible and fastest growing branch of artificial intelligence. Their objective is to capture the knowledge of an expert in a particular area, represent it in a modular, expandable structure, and transfer it to others.

Based on this definition, an expert system is a computer program that contains all of the following elements:

- An inference engine
- A knowledge base separate from the inference engine
- A data base
- An explanatory facility.

The *inference engine* is a computer program that draws logical conclusions, or inferences, from the expert knowledge contained in the knowledge base and the specific problem conditions contained in the data base. One discovery that made expert systems useful was the realization that, by separating the inference engine from the knowledge base, the inference engine can be used in any number of expert systems.

The *knowledge base* is a computer representation of the knowledge of the domain expert, expressed in one of several forms—for example, as rules. A knowledge base is specific to each expert system application.

The *data base* is a set of input data, pertinent to a particular problem, that is used by the inference engine to reach a conclusion. These data may be obtained from the user of the expert system, from instruments or sensors, from another computer data base, or from the output of another computer program.

The *explanatory facility* is a computer program that explains to the user why or how the expert system arrives at a particular conclusion. That is, if the user asks the expert system to explain its results, the explanatory facility determines the specific knowledge (from the knowledge base), combined with the user's input data, that led to this conclusion, and displays these to the user. The explanatory facility is one of the features that distinguishes an expert system from any other computer program.

A *rule base* is one kind of knowledge base, in which the domain expert's knowledge is expressed in the form of rules expressed in a natural language (e.g., English). These rules are typically of the form

IF   < Condition A is true >
        AND   < Condition B is true >

.
.
.

	AND	< Condition N is true >
THEN		< Conclusion X can be drawn >
ELSE		< Conclusion Y can be drawn >.

Not all expert systems are based on rules. Some are based on other knowledge-representational schemes, such as frames or objects. However, rules are the easiest for the novice to understand and are the basis for most available PC expert system development packages, or shells.

Figure 18-1 shows a user interacting with an expert system. The knowledge base in the computer has been created earlier by the domain expert. The user supplies the input data specific to his or her problem, or data may be extracted from a computer data base. The inference engine compares the user's input with the rules in the knowledge base to determine whether it can arrive at any conclusion. If not, the inference engine will ask the user for more information. The user may ask the inference engine why it needs this information, or why it has arrived at a certain conclusion, through the explanatory facility.

Figure 18-2 shows how an expert system is made. The domain expert who possesses the knowledge, interacts with a knowledge engineer, who is

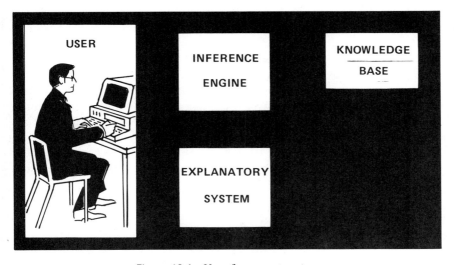

Figure 18-1 Use of an expert system.

Figure 18-2   Expert system development.

experienced in asking questions and converting the expert's answers int
rules. The rules resulting from this collaboration of domain expert an
knowledge engineer then form the knowledge base. Since the rules input t
the computer are derived from human judgments, they may be uncertai
insufficient, inconsistent, or even contradictory. Therefore, debugging th
knowledge base is a major task of the knowledge engineer.

After the knowledge base is complete, it should be tested in the field b
several users. An iterative process is always necessary to refine the know
edge base and to make the system reliable and easy to use.

Because the rules in the rule base are expressed in English (or somethin
like English), rules may be added over a period of time by the origina
domain expert or by others. This evolutionary improvement is characteris
tic of expert systems, as compared to other types of computer program
based on numerical algorithms. Of course, if many domain experts cor
tribute their knowledge, over time the expert system may become mor
capable than any one of the experts who contributed to its development

SYSTEM METHODOLOGY

Expert System Shells

Before 1985, it was generally believed that the development and use o
expert systems required specialized computer languages, such as LISP
and specialized computer hardware, such as LISP machines. With th

advent of personal computers (PCs) this has changed, and today many commercial expert systems shells are available, some for only a few hundred dollars.

An expert system shell is a computer program that combines an inference engine, an explanatory facility, and knowledge base development tools. A shell is essentially an expert system without a knowledge base, which must be created by the expert system developers. A shell can be applied to a number of different knowledge bases to create different expert systems. That is, different knowledge bases can be developed without the necessity of rewriting the inference mechanism, explanatory facility, and other functions each time.

An analogy with PC spreadsheets may be useful to those familiar with these programs. The expert system shell is analogous to the spreadsheet program as it comes from the box; it has no knowledge in it. The knowledge base is analogous to the rules or macros that the spreadsheet developer inputs to the spreadsheet program to define how to solve a particular type of problem. The data base is analogous to the data that the spreadsheet user inputs to the spreadsheet program to solve a specific problem. The explanatory facility has no analog in the spreadsheet program.

Thus, a primary factor opening up process and chemical plants to expert systems applications is the proliferation of general-purpose process plant computers, PCs, and inexpensive expert system shells. These computers at chemical plant sites offer the opportunity to deliver expert systems to the ultimate users.

System Development

With the realization that PC-based expert systems could solve significant problems, the commercial software market has become saturated with expert system building tools. Many people now have experience with microcomputers, and especially with spreadsheet programs. PC-based expert system shells often appear superficially to be as easy to use as spreadsheet programs. This may be a misconception. Many people have had prior experience with the manual solution of spreadsheets before learning to use microcomputers, but few have had prior experience with knowledge engineering and knowledge representation.

With expert system shells so easy to buy and apparently easy to use, should a person without experience undertake to build an expert system for personal use? There are a number of potential pitfalls in doing so. The selection of the proper form of knowledge representation requires experience and training. The selection of the best expert system tool for a particular application is complicated by the differences between conventional software and expert systems. Issues such as problem solving strategy, knowledge representation method, development environment, and ability to interact with existing systems must be considered, in addition to all of the standard interface, cost, and licensing issues associated with computer software.

These difficulties are why many expert systems are begun but not finished. Many people without training, buy a shell and start to build an expert system and then become discouraged when it does not work.

To address these issues for internal expert system development, it is necessary to produce guidelines for the development of expert systems for use by domain experts and knowledge engineers. These guidelines should explain the various inference mechanisms, the expression of knowledge in the form of rules, the selection of shells, and related subjects. A document such as this is desirable if personnel without experience in knowledge engineering are to develop expert systems with commercial shells.

Because few expert systems are available commercially, many people in the chemical and process industry will want to create their own expert systems. In such cases, to get the first expert system started on the right track, one may obtain the services of a consultant or take one of the short courses in expert systems available from universities, commercial educational services, or consultants.

Diagnosis of Problems

As noted above, the first expert system, MYCIN, was a diagnostic system, and expert systems for medical diagnosis are constantly being developed and improved. Diagnosis problems are well suited to solution by rule-based expert systems, because experienced domain experts familiar with expert systems can readily learn to express their diagnostic heuristics in the form of rules. However, domain experts who are not experienced in expert systems development may find it more natural to describe how a system works, rather than how it does not work. This approach will not generate a successful diagnostic system. Also, domain experts may find it natural to describe diagnostic rules in the form

IF   < Fault X is the cause of the problem >
THEN   < Symptoms A, B, and C will be seen >

However, this will not work either, because inference engines do not reason from causes to facts, but rather from facts to conclusions. Therefore, diagnostic rules should be expressed in the form

IF   < Symptom A is present >

```
        AND      < Symptom B is present >
        AND      < Symptom C is present >
THEN             < Fault X is the cause of the problem >
```

(Note that the last two rules are not necessarily equivalent.)

Rule-based expert systems may be incomplete; that is, the domain expert may not think of heuristic rules to cover every possible condition. This incompleteness may be resolved by the ELSE term in the IF . . . THEN . . . ELSE rule structure. In most cases, formal completeness will not be achievable, because of the heuristic nature of the diagnostic rules. Often, the domain expert may not recognize that his or her rule base is incomplete, because the rules cover the majority of the situations, and those not covered have a low probability of occurrence. For these reasons, it is highly desirable to have a domain expert available to back up the expert system if a situation arises for which the rule base is inadequate. These simple examples show why the assistance of a knowledge engineer may be necessary to help domain experts express their knowledge in the form of rules.

Expert systems have been developed for a wide range of systems and equipment diagnosis problems. Each step in the flow of diagnostic reasoning follows logically from the observed symptoms, to the probable causes of the symptoms, to remedies for the causes.

The inference engine determines the order in which questions are asked to the user, and the questions typically require either a Yes-No answer or a multiple-choice response; the user does not input long strings of English text. PC-based shells typically do not have the ability to interpret such text. Moreover, chemical and process plant personnel may not be good typists, so extensive text input would be undesirable. In most cases, the user need only enter the line number of the appropriate response.

When the inference engine has obtained enough data to determine the possible causes of the problem, the conclusions are presented to the user. The expert system may conclude one or more possible causes of the problem and may ask the user to verify whether the conclusion is in fact the cause of the problem. If it is, the expert system will advise the user on how to remedy the condition. If at any time the user does not understand the terminology, the expert system will provide an explanation or a tutorial on unfamiliar technical terms.

Chemical and Process Plants

Expert systems are well suited for providing advice to operators of process and chemical plants. Plant personnel who are not expert in some specialized field can solve problems for which they previously required the assistance of a senior person or a consultant. The knowledge of the expert is thus made available to many individuals at many locations, at any time of the day or night, and is not lost when the expert retires or changes positions. At present, most expert systems function as advisors to plant personnel who retain the ultimate decision-making authority.

REAL-TIME EXPERT SYSTEMS

Plant control systems are designed to control preset physical parameters in the plant. They are not designed to diagnose problems or to interpret data from a global plant perspective. Consequently, some of the problems that occur in plants, such as mechanical equipment malfunctions or valve failures, cannot be detected or prevented by the control system. Diagnosis and interpretation are the responsibility of the plant operators. The ability of expert systems technology to deal with issues diagnosis, interpretation, and prediction presents clear opportunities for augmenting existing control systems with real-time expert systems.

The benefit of implementing real-time expert systems in conjunction with process control systems is that plant operators now have the added dimension of expert interpretation of process and control data at their disposal. Expert systems reason about and provide advice regarding some of the issues to which conventional control systems are not well suited.

The development and implementation issues associated with real-time expert systems are similar to, but broader than, those for off-line interactive expert systems. Applications should be selected with the intention of providing expertise in areas where it was previously available only through the intervention of a human expert. Of course, human experts must be available to provide the knowledge base initially.

The development of real-time expert systems requires that considerable attention be paid to the hardware environment, because of the need to interface with existing process control computers and data acquisition systems, and because of the need for multiple processes to run concurrently.

Some of the operations- and control-related activities that are particularly well suited to expert systems applications are

- Equipment and process monitoring, with trend analysis, identification of spurious data, and identification of the causes of changes in process and control parameters
- Prediction of process conditions and product quality, based on evaluation of current plant parameters

- Integrated process control, combining real-time data, laboratory test data, and operator-entered data
- Intelligent alarming and establishment of alarm priorities
- Plant scheduling and process optimization.

The characteristics of expert systems that have made them attractive for off-line applications are also important for real-time applications. For ease of development by plant experts, construction of the knowledge base should not require knowledge of computer programming. Expert system development tools (shells) should be easy to use directly by the domain experts. The end user should be able to obtain explanations of all conclusions and recommendations, and any user interaction should be in natural language.

The types of plant control problems that can be solved include (but are not limited to)

- Instrument drifts and failures
- Mechanical equipment failures
- On-line analyzer calibration drifts
- Process-specific problems that may go undetected by the control system.

To accomplish these objectives, it is necessary to provide the expert systems with an understanding of both the manufacturing process and the plant control system. Knowledge of the process is embodied in rules governing the relationships between process parameters and describing the behavior of these parameters with respect to changes in the plant. Knowledge of the control system is contained in rules describing the behavior of the control system with regard to the process and the intended response of the controls to various process changes.

The development of expert systems involves representation of the heuristic rules that experienced operators and engineers would use to diagnose, or predict problems, in a form suitable to the inference mechanism. These rules may be specific to the particular plant, generic to the manufacturing process, or applicable to control systems in general.

Because many diverse functions are involved, covering the knowledge of many human experts, more than one expert system is required. The real-time expert system architecture must be specifically designed to accommodate multiple systems operating concurrently.

The architecture most suitable to the requirements of real-time expert systems is a blackboard system. The blackboard is a software structure that makes information available to various programs (see Figure 18-3).

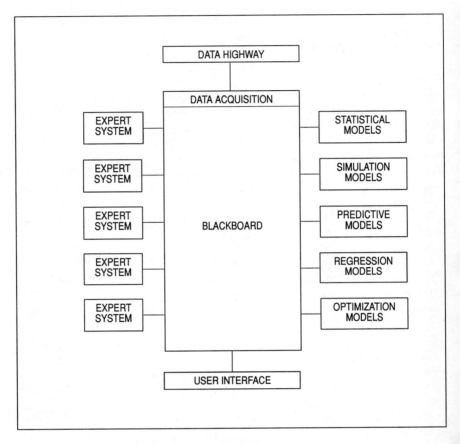

Figure 18-3 Blackboard system.

For example, the communications module updates the blackboard with new data, and the various expert systems use these data to make inferences about the state of the plant.

In this architecture, the expert systems can be developed and validated independently of each other and independently of the communications to the plant itself. The expert systems are simply given access to the information on the blackboard.

The expert systems may be developed using various commercial expert systems development shells. The expert systems operate in a similar fashion to the off-line advisors, with the exception that they obtain most of their information from the plant data highway, rather than from the end user. The expert systems are invoked internally, so they interact with the user only when they require information that cannot be derived from the plant data or when a conclusion has been reached and the plant operators are notified of the results. The expert system development shell should be selected for its ability to provide the explanatory and other features mentioned earlier, so that the operator can obtain explanations of the questions asked and the conclusions reached by the expert systems.

In addition to expert systems, other software, such as numerical simulation models for the plant, can be integrated into the system. Thus, the expert systems have access to model reference methods, statistical process control models, simulations, and so forth. By using both quantitative and qualitative techniques, balanced advice can be provided to the operators.

Rather than require a specific user interface, the system is flexible and can accommodate existing consoles within the plant (e.g., by tying into the graphics interface used for the control system) or can operate through a separate user console. The system provides the user with an interface that is easy to use, does not detract from the operator's normal activities, and provides immediate access to information.

To accomplish the requisite analysis, a continous monitoring of plant and controls data is required. The system should be designed to accommodate existing data communications facilities, including process or plant computers, and data highway interfaces. A direct data highway connection allows data to be examined and used by the expert systems as information becomes available.

The data are interpreted "on the fly" by data validation expert systems, and the interpretations are made available to the other expert systems. The other systems may be specifically designed to perform equipment diagnosis, instrument verification, and other functions.

The architecture can be applied to any plant or facility that has some form of distributed process control system. Application to an existing facility involves minor customization of the data acquisition software to conform to the available data highway, process control computer, or plant computer communications requirements. Plant operations and engineering personnel, together with knowledge engineers, develop the knowledge bases by tailoring the knowledge for the particular process and building the plant-specific knowledge bases.

Because of the independence of the expert systems from each other and from the software for data acquisition, the expert systems can be developed and implemented incrementally.

The types of plants for which this real time system is appropriate include

• Continuous-process chemical manufacturing plants

- Batch-process chemical manufacturing plants
- Petrochemical plants
- Industrial manufacturing plants
- Power plants
- Pulp and paper mills
- Pharmaceutical plants
- Food and beverage processing plants
- Wastewater treatment plants
- Water treatment plants.

RISK ASSESSMENT

The evaluation of risks requires a high degree of expertise. The construction of event trees, fault trees, and other analytical methods for risk assessment was discussed in earlier chapters. However, in practical cases, the information needed to construct fault trees may not be available, or the cost of these analytical methods may be too great to permit their use. An alternative is the use of expert systems, by which the heuristic knowledge of personnel experienced in risk evaluation can be readily put to use.

One example of a personal computer expert system in this area is the Technical Specifications Advisor[6]. Technical specifications are plant operating rules that govern the availability and operability of a plant's systems and equipment. Technical specifications may apply to safety considerations, environmental discharges, or other events that must be avoided. If a technical specification is violated, the violated condition must be restored to operability, or the system or even the plant may have to be shut down within a specified period of time.

Technical specifications are often complex and can be difficult to interpret and implement. Personnel who understand the technical specifications may be unavailable when needed, leading to unnecessary outages, or worse, failure to shut down when a shutdown is required. The Technical Specifications Advisor is an expert system that compares the actual or proposed plant conditions to the technical specifications and advises plant personnel when a shutdown is required or when equipment can be removed from service for maintenance or inspection without taking critical plant systems out of service.

For example, a plant safety train may require certain equipment to be operable. By inspection of the applicable piping and instrumentation diagrams, wiring diagrams, and other information, a rule may be derived such as the following:

IF   < The system is Fire Protection Train A >
    AND   < Valve A-V1 is Closed >

```
AND        < Breaker A-B1 is Closed >
  .
  .
  .
AND        < Pump A-P1 is Operable >
```
THEN   < Fire Protection Train A is Operable >
ELSE   < Fire Protection Train A is Inoperable >.

The status of the items of equipment listed in the condition part of the rule—namely, valve A-V1, breaker A-B1, pump A-P1, and so on—may be determined by querying the user or may be taken from the plant tagout system or data base of equipment status. This data base may be queried through a real-time system, as described earlier.

Given the status of Train A, the status of the complete fire protection system can be determined through additional rules, such as

```
IF               < The system is Fire Protection Train B >
    AND          < Valve B-V1 is Closed >
    AND          < Breaker B-B1 is Closed >
      .
      .
      .
    AND          < Pump B-P1 is Operable >
```
THEN   < Fire Protection Train B is Operable >
ELSE   < Fire Protection Train B is Inoperable >.
IF   < Service Water Supply Header Valve S-V21 is Operable >
    AND   < Service Water Supply Valve S-V23 is Fully Open >
    AND   < Service Water Discharge Valve S-V13 is Fully Open >
    AND   < Fire Protection Train A is Operable >
THEN   < Fire Protection Water Train A is Available >
ELSE   < Fire Protection Water Train A is Not Available >.
IF   < Service Water Supply Header Valve S-V22 is Operable >
    AND   < Service Water Supply Valve S-V24 is Fully Open >
    AND   < Service Water Discharge Valve S-V14 is Fully Open >
    AND   < Fire Protection Train B is Operable >
```
```

THEN   < Fire Protection Water Train B is Avail-
                          able >
ELSE   < Fire Protection Water Train B is Not Avail-
                          able >.
IF   < Fire Protection Water Train A is Not Avail-
                          able >
            AND   < Fire Protection Water Train B is Not Avail-
                          able >
THEN   < Fire Protection Water Flow is Not Avail-
                          able >
ELSE   < Fire Protection Water Flow is Available >.

Note that the last rule states that fire protection water is available if Train A or Train B, or both, are operable.

Suppose, in this simple example, that plant maintenance personnel want to remove pump A-P1 from service for inspection of the seals. At the same time, other personnel have opened breaker B-B1. Use of the Technical Specifications Advisor would indicate that fire protection water is available, even though Train B is inoperable, because Train A is operable. The postulated removal of Pump A-P1 would render Train A inoperable, however, and no fire protection water would be available. Thus, in this case, the inspection of Pump A-P1 must be postponed until Breaker B-B1 is closed, to avoid placing the plant at risk of fire.

Probabilistic Risk Assessment

Expert systems can also be used to evaluate intangible or qualitative factors that are used, for example, in quantitative risk assessment programs such as probabilistic risk assessment. In this case, the user inputs a subjective assessment of each of these factors, and the expert system generates a numerical estimate of the risk associated with each factor.

For example, expert systems can be used to evaluate failure probabilities, as in the following rule:

IF   < Pipe Material is type Stainless >
            AND   < Pipe Diameter is greater than 12 inches >
            AND   < Fluid Temperature is greater than 200°F >
              .
              .
              .
            AND   < Pipe Pressure is less than 1,000 psi >

THEN   < Pipe Rupture Probability is 0.00000001 >
ELSE   < Pipe Rupture Probability is 0.0000001 >

Expert systems can also be used to evaluate subjective factors, such as plant operator capabilities, as in the following rules:

IF   < Plant Operators Score on the written test is greater than 70 >
    AND < Plant Operators have Education at least High School Diploma >
    AND < Plant Operators have Experience more than 2 years >

    AND < Plant Operators' Morale is High >
THEN   < Plant Operators' Qualifications are High >
ELSE   < Plant Operators' Qualifications are Low >
IF   < Plant Operators' Qualifications are High >
    AND < Plant Operations are Not Highly Critical >
THEN   < Probability of Operator Failure is 0.01 >
ELSE   < Probability of Operator Failure is 0.25 >

Note that, in the last example, the responses to the conditions concerning the test scores, education, and experience are objective values that may be obtained from some data base, whereas the responses to the conditions on operators' morale and degree of criticality of the plant operations are subjective and must be provided by the user based on his or her judgment and assessment of the situation. In all cases, however, the result of the rules is a quantitative probability of failure that may be used in subsequent risk assessment calculations.

In a similar manner, expert systems can be used to interpret the quantitative results of numerical risk assessment calculations. For example, a risk assessment program may compute the probability of a toxic chemical release. Rules to evaluate this result might look like the following:

IF   < Probability of Chemical Release is greater than 0.5 >
    AND < Toxicity of Chemical Release is High >
    .
    .
    .

	AND	< Expected Morbidity at Plant Boundary is greater than 1 >
THEN		< Install additional redundant safety systems >
	AND	< Provide protective gear to all personnel >
	AND	< Install warning devices at plant boundary >

In many cases, a quantitative fault tree or event tree for a probabilistic risk assessment may not be available, either because the system is not understood well enough to create a fault tree, or because the analytical effort is too expensive and time-consuming. In such cases, a qualitative assessment can be performed using expert systems techniques.

In a typical rule, the conditions are related by a logical And. A risk assessment rule may then be stated in probabilistic form, if all the conditions are statistically independent, as

IF		< The probability that Condition A is true is P_A >
	AND	< The probability that Condition B is true is P_B >
	.	
	.	
	.	
	AND	< The probability that Condition N is true is P_N >
THEN		< The probability that Conclusion X is true is $P_A P_B \ldots P_N$ >

If the probabilities of all the conditions, P_A, P_B, . . ., P_N, are known, then the probability of the truth of the conclusion can be computed from the product of the probabilities.

Suppose that quantitative probabilities cannot be determined for various events. Instead, probabilities are assigned one of two qualitative values: high or low. A high probability is assumed to be close to 1, and a low probability is assumed to be close to 0. The product of probabilities, all of which are high, is assumed to be high, but if one or more of the individual probabilities is low, then the product is Low. To illustrate, consider the Technical Specifications Advisor example given earlier, this time in qualitative probabilistic form:

| IF | | < The probability that Service Water Supply Header Valve S-V21 is Operable is High > |

AND   < The probability that Service Water Supply
        Valve S-V23 is Fully Open is High >
AND   < The probability that Service Water Dis-
        charge Valve S-V13 is Fully Open is High >
AND   < The probability that Fire Protection Train A
        is Operable is High >
THEN   < The probability that Fire Protection Water
        Train A is Available is High >
ELSE   < The probability that Fire Protection Water
        Train A is Available is Low >.

In this probabilistic rule, Fire Protection Train A has a high probability of being available only if all the conditions have a high probability; otherwise, it has a low probability of being available.

The user of the expert system would assign judgmental values to the probabilities. For example, if the probability that the header valve S-V21 is operable may be considered Low, then the probability that Train A is available will also be Low.

Similarly, fire protection water is available if Train A or Train B, or both, is available. By assumption, if the probability that Train A is available is low, then the probability that Train A is unavailable is high, and conversely. Hence, the combining rule becomes:

IF   < The probability that Fire Protection Water
        Train A is Available is Low >
AND   < The probability that Fire Protection Water
        Train B is Available is Low >
THEN   < The probability that Fire Protection Water
        Flow is Available is Low >
ELSE   < The probability that Fire Protection Water
        Flow is Available is High >.

Thus, if Train A has a low probability of being available, but Train B has a high probability of availability, then the probability that fire protection water is available will be high. This chain of qualitative probabilistic reasoning can be carried forward, as in the deterministic expert system, to conclude the risk that the plant or system under consideration will fail.

In this example, the deterministic expert system rules regarding the operability of Trains A and B and the availability of fire protection water were derived from the plant flow diagrams and the expert's knowledge of the plant, as described earlier. These rules can then be turned into quantitative or qualitative probabilistic rules, depending on the information available and the objectives of the system.

Confidence Factors

Another approach to risk assessment with expert systems uses confidence factors. Confidence factors are values assigned to each fact that describes the confidence or certainty that the fact is true. There are many different methodologies for computing confidence factors. Some systems assign confidence factors values between 0 and 1, as with probabilities, whereas other systems use values between -1 and $+1$, or between -10 and $+10$, and so on. In Bayesian methods, the confidence factors are treated as probabilities, and the computation of dependent confidence factors uses the calculus of probabilities. However, most systems do not equate confidence factors and probabilities.

In addition to the confidence factors assigned to facts, there may be confidence factors assigned to each rule. This rule confidence factor represents the domain expert's confidence that the rule is valid. Thus, given a rule, a confidence factor assigned to the rule, confidence factors for all the facts used in the conditions of the rule, and a method of calculation, it is possible to compute the confidence factor for the conclusion of the rule. If this conclusion is used in the condition parts of other rules, the confidence factor computed for this conclusion will be used to calculate the confidence factors associated with the conclusions of these subsequent rules. The method for combining confidence factors, or measure of belief, in the MYCIN program is given in Chapter 11 of *Rule-Based Expert Systems: The MYCIN Experiments of the Stanford Heuristic Programming Project,* by Buchanan and Shortliffe[4].

A complete description of the various methods for combining confidence factors is beyond the scope of this chapter. A simple method states that the confidence factor for the conclusion of a rule is the minimum of all the confidence factors for the facts in the conditions (assuming all conditions are ANDs) multiplied by the rule factor. To illustrate, consider again the fire protection rules from the Technical Specifications Advisor:

IF		< Service Water Supply Header Valve S-V21 is Operable >
	AND	< Service Water Supply Valve S-V23 is Fully Open >
	AND	< Service Water Discharge Valve S-V13 is Fully Open >

	AND	< Fire Protection Train A is Operable >
THEN		< Fire Protection Water Train A is Available >
ELSE		< Fire Protection Water Train A is Not Available >.

Suppose the confidence factors are

Service Water Supply Header Valve S-V21 is Operable	0.60
Service Water Supply Valve S-V23 is Fully Open	0.70
Service Water Discharge Valve S-V13 is Fully Open	0.80
Fire Protection Train A is Operable	0.95
(Determined from the conclusion of the previous rule)	
Rule factor	0.90

Then the confidence factor for the conclusion, that Train A is available, is the minimum of all the certainty factors for the facts in the conditions, or 0.60, multiplied by the rule factor, 0.90, which yields 0.54. It is often, but not always, assumed that the confidence factor in the negative result, here that Train A is not available, is the complement of this, or 0.46.

Suppose that the confidence factor for Train B unavailability, from a similar rule, is 0.44. Then, by the rule

IF		< Fire Protection Water Train A is Not Available >
	AND	< Fire Protection Water Train B is Not Available >
THEN		< Fire Protection Water Flow is Not Available >
ELSE		< Fire Protection Water Flow is Available >.

the confidence factor for the availability of fire protection water would be 0.56; the confidence factor for unavailability of water would be 0.44. This may be interpreted as, in some sense, an expression of the risk of having no water to fight fires.

By the appropriate use of confidence factors, expert systems can be built to determine the risk for any type of event associated with a chemical or process plant.

A more complete treatment of the subject of confidence factors and degrees of belief, the Dempster-Shafer theory of evidence, is given in the classic book by Shafer, A *Mathematical Theory of Evidence*[7].

CRISIS MANAGEMENT

Crisis management is required when risk prevention fails and an accident occurs or is imminent. Crisis management is made difficult by the need for fast, correct responses under conditions of great uncertainty and pressure. Mistakes can be costly and dangerous to plant personnel, the public, and the environment. Almost invariably, crises occur without warning, at the wrong time, at the wrong place, and with the wrong people on the scene. The absence of knowledgeable experts can turn a minor problem into a major incident. Consequently, the application of expert systems to crisis management can have significant benefits.

The existence of crisis management expert systems can also affect risk assessment as well. Any risk assessment process should consider the effects of human error. In fact, many of the most visible incidents in plants in recent times (Three Mile Island, Chernobyl, Bhopal) can be traced to human failures rather than equipment failures. If expert systems were available at points at which human decisions were required, whether for normal operations, problem diagnosis, or crisis management, the probabiltiy of human failure might be substantially reduced. Thus, expert systems could reduce risk and should therefore be taken into account in the risk assessment. Conversely, risk assessment methods can identify the points at which human failure could be most serious, and thus identify areas in which expert systems should be developed and applied.

Some of the characteristics of crisis management that make it suitable for expert systems are

- Experience and knowledge of relatively few experts
- The infrequent and random character of the events, which most often occur when experts are not present
- A high degree of uncertainty in the information available
- A high-pressure, time-critical environment in which immediate action is preferable to time delay
- The existence of numerous regulations, which may be difficult to access and to interpret
- The need for immediate, correct decisions and action by on-the-spot operations personnel, who may not be familiar with the nature of the crisis and the means for resolving it
- Attention by the media and by political figures seeking to take advantage of any indecision or uncertainty on the part of crisis managers.

Spill Management

One area of crisis management that has received much attention from the viewpoint of expert systems is chemical spill or release management. For example, Oak Ridge National Laboratory developed a knowledge-based system for oil and chemical spill management in 1980[8, 9]. An expert system for chemical spill or release management may have the following components or functions, which are generally characteristic of a broad class of crisis management systems:

- Detection
- Characterization
    Material
    Source
    Magnitude
    Destination
    Hazard assessment
- Response
- Notification
- Analysis.

Detection

Chemical releases or spills may be discovered by workers or detected by instruments. In the case of discovery by workers, procedures should be in place for the immediate notification of appropriate management personnel, as delays can be extremely costly. Detection by instruments, either special-purpose instruments or normal process instrumentation, can be augmented by real-time expert systems, as discussed earlier. Such real-time expert systems can detect releases earlier and in smaller quantities and can initiate the identification and response process automatically, giving more time to assess and respond to the situation.

Characterization

Identification of the material   Identification of the material and its source are the two most immediate problems when a release has been detected. The identification of the toxicity of the material may have priority even over identification of the source, because special equipment (respirators, protective suits, etc.) may be required to protect workers sent to stop the flow of material. There have been numerous instances of workers overcome by gases as they tried to shut off the flow of the discharge without protective gear. Conversely, in California a major highway was closed to traffic when a shipment labeled "Ferric Oxide" was spilled, and no one present knew this to be common, ordinary iron rust.

Special-purpose instruments and real-time process control expert systems may be able to detect and identify the material in very small concentrations. Releases discovered by personnel can be identified by interactive expert systems based on color, odor, viscosity, opacity, container labels, shipping lists, and other information. Such expert systems must operate on incomplete or even conflicting information obtained from various observers. In many cases, a preliminary characterization of the type of material is needed to assess the degree of hazard or toxicity as rapidly as possible, with a final determination based on additional evidence. Expert systems based on confidence factors may be useful here. In plants, the expert system should be tied into a data base of known chemicals at the site, to exclude unlikely candidates.

Identification of the source If the material is not acutely toxic or hazardous, identification of the source of the discharge may be the highest priority to stop the release before it becomes larger. In some cases, the source may be known from the identification of the material, using a data base of plant sources available to the expert system. In other cases, it may be necessary to backtrack through a system, process, discharge line, or sewage line to identify the source of the release. This can be readily accomplished if the expert system contains a logical map or diagram of the site or plant process[7].

Identification of the magnitude of the release The type and degree of the response may depend upon the magnitude of the release, as well as on the type of material. A data base of quantities of materials on hand, in storage, or in production is very helpful to the expert system in this assessment. In cases of visual detection, the expert system can help to assess the size of a spill from visual indicators, such as area covered and apparent depth.

Identification of the destination of the discharge This characterization is necessary to warn persons downwind or downstream and to take actions to contain the release. Based on the estimated magnitude of the release, environmental conditions (wind speed and direction, etc.), and internal maps, diagrams, and networks, as in the case of source identification, an expert system can identify the probable destination and extent of the release.

Hazard assessment Using all the information available, including the identification of material, source, magnitude, and destination, an expert system can assess the degree of hazard. This assessment is similar to the process of risk assessment, discussed earlier.

Response determination

Based on the foregoing characterizations, the expert system can provide advice on the responses to be taken to terminate, contain, alleviate, and clean up the release. Various federal, state, and local regulations specify actions that must be taken. Individual companies and plants will also have organization-specific actions to be followed, depending on circumstances. An expert system could identify the most appropriate and the legally required actions, and assure that none are omitted, that they are performed in the correct sequence, and that the organizational response meets all safety requirements and regulations.

Notification

Government regulations at all levels specify complex notification requirements depending on the type, location, and magnitude of chemical releases. Financial liability and public credibility may depend upon making these prescribed notifications in a timely manner. Many plant personnel are not familiar with such requirements, as the need is infrequent, and procedural violations often occur. An expert system can identify the required notification process and can initiate transmission of the required notification and documentation automatically, if so desired.

Analysis

No job is finished until the paperwork is done. After a chemical spill, an analysis of the incident is always required. This may include an analysis of the causes of the discharge, an evaluation of the performance of the crisis management process, and an assessment of ways to prevent future occurrences. An expert system can assist in such after-the-fact evaluations, and in addition can itself embody the results of the evaluation, in the form of additional or improved rules to be used in any future situation. Thus, the expert system can learn from experience.

As shown in the preceding description, an expert system for crisis management is based on

- Heuristic rules derived from experience in previous crises and the knowledge of crisis management experts

- Regulations
- Access to data bases of plant chemicals, inventories, processes, and related information on a timely basis.

Each expert system must be tailored to the specific situation and conditions of a plant, a site, or an environment. Such expert systems should be tested, through simulated crises, to make sure that they perform under the stress of real situations. These crisis management expert systems can also be used to train plant operations and crisis management personnel, so that real incidents can be minimized or prevented.

CONSTRAINT-BASED REASONING

The heuristic rules used in rule-based expert systems are essentially short-cuts, representing the distillation of the experience and judgment of one or more domain experts. Expert systems based on heuristic rules are appropriate when these experienced domain experts exist. However, in many situations, no domain experts are available. In such cases, constraint-based systems can provide an alternate solution.

Constraint-based reasoning is essentially a qualitative simulation of a process based on constraints. Constraints are rules that are based on physical principles rather than on the heuristic judgment of experts. Consequently, constraint-based systems can be generated even in the absence of experts. Such a situation would occur in the design of a new plant or a retrofit to an existing plant. For example, a constraint-based system for qualitative simulation of a proposed plant system can be automatically generated from a process flow diagram.

The objective of a constraint-based system is to predict the behavior of a plant system, based on given inputs, in a manner similar to that used by a skilled engineer unfamiliar with the specific system but experienced with similar processes. That is, the analyst would reason from causes to effects in small steps, working from the given inputs through the components of the system to reach the conclusions. This reasoning would be based on the physical characteristics of the components and their connectivity, rather than on heuristics or shortcuts derived from experience with this particular system.

The constraint-based system essentially replicates this reasoning process, using the constraints to reason from step to step through the process. Because the constraints are derived from physical principles pertinent to each component of the plant system being simulated, they can be cataloged and filed in libraries for all the standard components of a plant, such as valves, pumps, and tanks. The generation of a constraint-based system for a particular process can be automated, requiring only the constraint libraries and the topology of the system in question.

CONCLUSIONS

The power of expert systems lies in their ability to separate knowledge about specific areas of application (the knowledge base) from general problem-solving knowledge (the inference engine). This separation is accomplished by the expert system shell.

Much of what is considered to be expertise is actually experience distilled in the form of heuristic rules. Experts are those people who have accumulated more rules, in a certain specific area of interest, than have non-experts. These heuristic rules, if expressed verbally, can be coded into expert systems that embody the superior problem-solving capabilities of the human experts.

The value of expert systems lies not in the replacement of experts by computer programs. There is no known instance of this ever occurring, because no expert system is as good as a human expert. The value of expert systems lies in their ability to replicate the performance of experts on judgmental tasks in situations in which there are not enough human experts, because they are retired, sick, on vacation, in another location, or otherwise not available.

In the proper situations, expert systems can improve productivity, reduce errors, and improve quality. Many expert system applications in chemical and process plants have proven to be valuable, but these have barely scratched the surface of the possibilities in this field. Process plant risk assessment is one area in which expert systems will prove to be very beneficial.

REFERENCES

1. Untitled news item, 1988. *Manufacturing Engineering,* 103(1), p. 21.
2. Feigenbaum, E. A., McCorduck, P., and Nii, H. P. 1988. *The Rise of the Expert Corporation.* New York: Random House Times Books.
3. Barr, A., and Feigenbaum, E. A. 1981. *The Handbook of Artificial Intelligence.* Los Altos, CA. William Kaufmann, Inc.
4. Buchanan, B. G., and Shortliffe, E. H. 1984. *Rule-Based Expert Systems: The MYCIN Experiments of the Stanford Heuristic Programming Project.* Reading, MA: Addison-Wesley.
5. *Report of the research briefing panel on computers in design and manufacturing.* 1983. National Academy of Sciences—National Academy of Engineering, p. 60.

6. Stone & Webster Engineering Corporation. 1986. *Expert Systems Services.* Boston, MA.
7. Shafer, G. 1976. *A Mathematical Theory of Evidence.* Princeton, NJ: Princeton University Press.
8. Oakes, T. W., Bird, J. C., Shank, K. E., Kelly, B. A., Clark, B. R., and Rodgers, F. 1980. *Waste Oil Management at Oak Ridge National Laboratory.* ORNL/TM-7712.
9. Johnson, C. K., and Jordan, S. R. 1983. Emergency mangement of inland oil and hazardous chemical spills: A case study in knowledge engineering. In *Building Expert Systems.* Frederick Hayes-Roth, Donald A. Waterman, and Douglas B. Lenat, Eds. pp. 349-97. Reading, MA: Addison-Wesley.

19

Directions in Legislation and Regulation

Joseph J. Cramer

For many years, chemical and petrochemical industry emissions have been regulated, and permits have been required for the discharge of sulfur oxides, nitrogen oxides, and certain hydrocarbons, Similarly, liquid discharges have also had to comply with federal and state permitting requirements, and the industry has dealt with local or state regulations that attempt to control nuisances or public endangerment. Nevertheless, because of the multitude of chemicals and the diversity of their properties, many routine releases have not been specifically regulated.

The nature of a process facility is to "process" chemicals. Hazardous chemicals are not intended to be routinely released; however, releases do occur accidentally, and sometimes safety systems are designed to vent or release chemicals to protect against more catastrophic events, such as explosions or fires. These accidental releases are sometimes addressed by general enviornmental regulations designed to ensure public health and safety; the implementation of these regulations might require postrelease notification and reporting, and possibly result in fines or penalties. However, until quite recently the plant operations and management practices that control the potential for such releases have not been specifically regulated.

The industry has felt very strongly that responsibility for the design and operation of chemical facilities was its own, because it involved highly specialized, often proprietary, knowledge and experience. Regulatory agencies were not thought to have the expertise or resources to be able to be significantly involved in the process. Many in the industry also believed that it should not be the business or concern of regulators, because the industry's record for safety was exemplary.

Much of this thinking was shaken by the tragedy at Bhopal. The deaths of 3,000 or more people from the acute toxicity of the methyl isocyanate

(MIC) releases, and the injury of tens of thousands more, had a dramatic effect on the U.S. public, and on state and federal legislators and regulators. Today's rapid transportation and communication networks made the events seem immediate and personal. The media attention focused on the Bhopal plant's owners, Union Carbide, and on the industry in general. The tragedy at Bhopal also magnified the public's awareness of, and reaction to, lesser accidents. An incident involving a release from another Union Carbide facility that occurred in 1985 in Institute, West Virginia, although unrelated to MIC and with no apparent long-term consequences, served to further increase mounting regulatory pressures.

The circumstances of the accident at Bhopal made it appear that improved safety or risk management programs could have prevented or greatly mitigated the consequences of the accident. Previously, the concept of risk management had become well established in both the aerospace and nuclear industries, so it was not surprising to see numerous pieces of proposed legislation appear quickly at both the state and federal levels. The status of these actions and likely future directions are reviewed in the following sections and are later summarized.

Some recent chemical plant safety milestones are:
December 1984 Bhopal catastrophe—created strong initial driving force for accident prevention and risk management efforts
August 1985 Institute West Virginia accident—reinforced increasing level of concern in the United States
October 1986 Passage of SARA Title III—served to establish chemical safety as a continuing concern
Early 1986 Adoption of initial state risk management legislation.

A chronology of developing risk management legislation and regulations is as follows:
January 1986 Passage of New Jersey Toxic Catastrophe Prevention Act (TCPA)
• First state legislation that dictates to some degree how to operate facility to ensure safe operation
• Reguires risk management programs
• For those found lacking; mandates third-party audits
• Initially covered 11 chemicals
1986 Adoption of California's Risk Management and Prevention Plan Legislation (AB3777)
• Also requires risk management programs
• Lacks detailed implementing regulations
June 1988 Adoption of *Final TCPA resolutions*

• Additional chemicals added to list to bring total to 103 chemicals
July 1988 Adoption of Delaware's Extremely Hazardous Substances Risk Management Act
December 1988 Publication of Recommendations for Process Hazards Management of Substances with Catastrophic Potential (Organization Resources Counselors—OSHA)
July 1989 Publication of California Guidance for RMPP
• Detailed and intensive
September 1989 Adoption of Delaware implementing regulations
January 1990 Issuance of *API Recommended Practice 750—Management of Process Hazards.*

FEDERAL ACTION

The principal federal response through 1989 had been the passage of The Emergency Planning and Community Right-to-Know Act of 1986. This Act, which was passed as Title III of the Superfund Amendments and Reauthorization Act (SARA) of 1986, required action from industry and from state and local governments. The act was adopted primarily as a compromise between those desiring stronger, more active risk management legislation, and those who felt that no major legislation was required. Stated broadly, the act required the preparation of local emergency response plans for dealing with a number of extremely hazardous substances, and it established procedures and mechanisms for reporting accidental releases and for releasing information on the inventory, location, routine release, and properties of numerous hazardous chemicals. The act was designed to require a greater degree of interaction, cooperation, and planning between local and state governments, wide segments of the public, and industry. This was to be accomplished by having the local emergency response plans prepared by local emergency planning committees, which were to have broad public representation, as well as representation from the facilities handling the specified hazardous chemicals. A state emergency planning commission was to oversee the work of the local committees.

The emergency response features of the act were similar to the Chemical Emergency Preparedness Program of the Environmental Protection Agency (EPA), and to the Chemical Manufacturers Association (CMA) Community Awareness and Emergency Response (CAER) program. Both of these programs represented voluntary efforts to encourage more cooperative planning and the open exchange of information. The CMA program is aimed primarily at industry, while the EPA's program targets local communities and governments.

The act's implementation is gradually creating a greater public awareness of the quantities of chemicals and potential hazards present in our society. This is resulting in increasing pressure on industry to adopt effective and more visible risk management programs. The act also called for the EPA to initiate a comprehensive report to Congress on the review of emergency systems. While the resulting report did not specifically recommend the adoption of formal risk management programs or risk management legislation, it did recommend the strengthening of a number of the elements that make up a risk management program.

The most significant long-term effect of SARA Title III is likely to be a greater public awareness of the wide variety of chemicals with greatly varying properties that are present throughout the country. This greater public awareness, not only of the chemical threat, but also of the possible controls or management responses, should build a pressure and consensus for more active risk management. This has already manifested itself in the adoption of stronger voluntary programs, and the enactment of additional state and federal regulatory programs.

In fact, the recently-passed Clean Air Act Amendments of 1990 include provisons that will require the adoption of regulations for the prevention and detection of accidental chemical releases. These provisions will affect a minimum of 100 chemicals and are to be promulgated by late 1993. The regulations must include a provision for risk management plans covering all the elements addressed in Chapter 1. The Clean Air Act Amendments also call for the Occupational Safety and Health Administration to adopt regulations to protect against workplace hazards resulting from accidental chemical releases. These requirements basically cover the same general elements addressed by the process safety management standards proposed by the Occupational Safety and Health Administration in July of 1990. It is possible that by the time that these regulations (29CFR Part 1910.119) are adopted, they will satisfy the Clean Air Act's requirements for workplace protection.

STATE INITIATIVES

Several states were not satisfied with the early pace of federal action and enacted programs of their own for meeting what they perceived as the toxic chemical threat. While several states adopted emergency planning or community right-to-know bills similar to SARA Title III, only three states have to date passed risk-management—type legislation: New Jersey, California, and Delaware. Of these, only New Jersey's Toxic Catastrophe and Prevention Act, and California's Acutely Hazardous Materials Risk Management legislation have been in existence long enough to have had some effect on the regulated community.

Although both New Jersey and California legislation requires many measures for risk management, there are differences in both the content and administration of the regulatory programs. It is not our intent to do a detailed comparison between these programs, because they are both still in the evolutionary phase. However, it is enlightening to touch on some of the key features in each program and to note the most significant differences and similarities, as summarized in the following list and in Table 19-1.

- New Jersey program—Penalties include heavy fines and criminal penalties.
- New Jersey's regulations are very prescriptive.
- New Jersey—Numerous third-party audits ordered.
- California—120 Administering agencies: police, fire and health departments.
- California's regulations are brief, but guidance is intensive.
- California program initiation is stretched out because of number of agencies and requirements to request RMPP.
- Both California and New Jersey act as ''de facto'' permitting programs for new facilities.

The most striking administrative difference between the two programs is that New Jersey's program is centrally administered by the New Jersey Department of Environmental Protection (NJDEP), while California's program is locally administered. Risk management programs in New Jersey must be approved by the NJDEP and are to be inspected on a periodic basis after initial approval. Those companies failing to gain approval of a program must undergo an extensive third-party safety and risk evaluation at their own cost, which eventually should lead to an approved risk management program. A novel feature of this requirement is that the NJDEP selects the consultant to do the study from three consultants nominated by the registrant. The NJDEP has added several senior staff members with chemical industry experience to handle the extensive responsibilities assigned to them by the New Jersey legislation. Fees and fines associated with the program are intended to make it self-supporting.

California's program is to be administered by approximately 120 local administering agencies, which can be local fire departments, health or environmental agencies, and police departments. Also, while both states have published lists of chemicals and thresholds of applicability, New Jersey automatically requires each affected party to have a risk management program for each applicable chemical, while in California, a risk management and prevention plan is required only after an administering agency formally requires one.

Table 19-1   Highlights of Regulatory Programs

Program	Law or Regulation	Direction	Chemicals	Threshold Quantitities	RMP Required Automatically
New Jersey	Yes	Centralized	11 Initially 93 Added. More will probably be added	10 To 34,000 LB. (Based on IDLH, LD_{LO}, $LC_{50/10}$, TLV and STEL)	Yes
California	Yes	Local	366 (Federal list)	500 LB 55 GAL 200 Cu Ft.	No
Delaware	Yes	Centralized	181 (3 Categories)	Varies toxicity (ATC/SHI) explosives flammable & combustible	Yes
ORC[a]	Not yet	Either	96 Toxic chemicals	SHI>5,000	Not specified
API[b]	No	N/A	Unspecified	SHI>75,000	N/A

[a] *Organization Resources Counselors, Inc.*
[b] *American Petroleum Institute*

Substantively, New Jersey's program is, at least on the surface, more prescriptive and requires an eight-element program:

- Safety review
- Hazard analysis (risk assessment)
- Operating procedures
- Preventive maintenance program
- Operator training
- Accidental investigation procedures
- Emergency response program
- Audit program.

New Jersey has issued extensive regulations defining what constitutes acceptable elements. California's legislation covers the same general areas, but not as specifically and without detailed implementing regulations. However, the California program has detailed and quite stringent guidelines for use by the administering agencies in the implementation of the program. It is likely that many of the local administering agencies, lacking specific expertise, will apply those guidelines with little or no modification. The resulting program could, therefore, be just as extensive and prescribed as the one in New Jersey.

California's regulations currently cover several times the number of chemicals as those addressed by New Jersey's Toxic Catastrophe Prevention Act, but New Jersey's program is designed to be expanded and may well be increased to cover more than the 103 chemicals now listed. At this time, New Jersey regulates only chemicals that represent an atmospheric threat, but the legislation behind the regulations has no such limitations. In addition, inclusion or use of a chemical on California's list does not automatically trigger preparation of a plan, so the number of plans required in each state may be comparable.

It is also worth noting that both states have a de facto permit program that requires risk plans to be approved before the construction or operation of new units handling the regulated chemicals.

It will take some time to gauge the relative successes of the two administrative approaches. Because of the many other differences between California and New Jersey, it may never be completely possible to judge the merits of each approach. More importantly, each program requires the implementation of an extensive and very active chemical risk management program that definitely emphasizes prevention.

The program now being developed in Delaware to implement Delaware's Extremely Hazardous Substances Risk Management Act is following the same path as the programs in California and New Jersey. The legislation provides for at least an eight-element risk management program. The elements, while not corresponding exactly with those of either New Jersey or California, cover essentially the same ground. Delaware's program will be centrally administered by the Department of Natural Resources and Environmental Control. However, the role of the department in the administration of the legislation would appear to be more passive than that assigned to the NJDEP for the Toxic Catastrophe Prevention Act. The Delaware program represents a continuation of the trend toward state regulatory programs that aim at the prevention of accidental chemical releases.

ADDITIONAL FEDERAL ACTIONS

It is likely that before many more states adopt their own preventive risk management legislation, additional federal regulatory action will be taken. In late 1988, Organization Resources Counselors, Inc. (ORC) completed a report for the Occupational Safety and Health Administration (OSHA) entitled "Recommendations for Process Hazards Management of Substances with Catastrophic Potential." The report was commissioned to support OSHA in the preparation of revisions to its standards for hazardous materials (29 CFR Part 1910 Subpart H). The document was completed in the form of draft regulations complete with definitions, proposed chemicals to be covered, design basis references, hazard analysis methodologies and references, and guidelines for accident investigation. The program contains an 11-element risk management program that includes the same requirements as New Jersey's program, plus quality assurance and change procedure controls. As stated previously, OSHA has now issued draft regulations along these lines which may eventually also satisfy the requirements of the Clean Air Act Amendments of 1990. It is interesting that, although OSHA's primary goal is to protect workers, the type of program being considered is very similar to the programs of New Jersey, California, and Delaware, which are intended to protect the general public. Specific elements of each program are predicted in Table 19-2.

In addition to the OSHA activity, the Clean Air Act amendments also contain provisions requiring the periodic application of hazards analysis techniques to a number of specified chemicals. The intial list includes only 100 chemicals, but can be supplemented as required. This new legislation also calls for the creation of a National Chemical Safety Board, modeled after the Civil Aeronautics Board, which would have broad power to examine all serious chemical accidents and make recommendations to the EPA for possible mitigative action. Although the specific elements will not be clarified for several years, plans will be required, and the administrative mechanisms and authority to require additional or stronger risk management elements are in place.

Table 15-2 Required Program Elements

API 750	New Jersey	California	Delaware	ORC
Process safety information	Safety reviews	AHM accident description	Design standards review	Process safety information review
Process hazards analysis	Hazards analysis/risk assessment	Equipment description	Modification procedure	Process hazard analysis
Management of change	Standard operating procedures	Description of design, operating and maintenance controls	Hazard review	Management of change procedures (Process or Facility)
Operations procedures	Preventive maintenance program	Description of detection, monitoring, and automatic control systems	Operating instructions	Operating procedures
Safe work practices	Operator training programs	Risk reduction plan and schedule	Maintenance and inspection procedures	Training
Training	Accident investigation procedures	Auditing and inspection programs	Operator training programs	Indcident investigation
Assurance of the quality and mechanical integrity of critical equipment	Emergency response program	Record-keeping procedures	Incident investigation procedures	Compliance Review (audit)
Pre-startup Safety review	Audit program	RMPP should be based on a hazards and consequence analysis, which may include any of the following:	Auditing requirements	Equipment quality assurance
Emergency response and control		HAZOP		Mechanical integrity (maintenance)
Investigation of process-related incidents		Dispersion Analyses		Pre-startup safety review
Audits of process hazards management systems		Quantitative risk assessment		Emergency response and control
		Human reliability analysis		
		Seismic assessment		

INDUSTRY ACTION

In addition to the CMA programs, the principal industry response to Bhopal and regulatory initiatives has been the creation of the Center for Chemical Process Safety (CCPS) of the American Institute of Chemical Engineers. Sponsored by approximately 50 of the largest chemical, petroleum, pharmaceutical, industrial, engineering, and consulting companies, the center has put considerable effort into the preparation of various guidelines for the development of risk management programs. While of great value to individual companies implementing a risk management program, or to regulators, it is unlikely that this program or CMA's will blunt the gradually growing trend towards risk management regulation. Additionally, in January 1990, the American Petroleum Institute issued API Recommended Practice 750, Management of Process Hazards. This guideline uses an 11-part risk management program that corresponds almost exactly to the ORC program. Similar criteria for initiation are also advanced but are less specific on what should be included in each element.

CONCLUSIONS

The movement towards mandated risk management programs is likely to continue. At this time, action appears to be measured and cautious but reasonably steady. As in New Jersey, California, and Delaware, further legislative or administrative actions are being contemplated in New York and Illinois. The slow progression is probably fortunate, in that it is allowing many individual companies and groups such as CCPS and CMA to develop improved analytical and management tools and tecniques, that should ease the ultimate implementation of required programs.

Hopefully, the regulative programs adopted will be reasonable and feasible in both goal and approach. The slow but steady progress toward mandatory risk assessment could be a headlong rush if any major accident were to capture the public attention or raise its present level of concern.

Acronyms

ACGIH	American Conference of Governmental Industrial Hygienists
AGA	American Gas Association
AHM	Acutely Hazardous Material
AIChE	American Institute of Chemical Engineers
AIHA	American Industrial Hygiene Association
APELL	Awareness and Preparedness for Emergency at the Local Level
API	American Petroleum Institute
ATC	Acute Toxicity Concentration
BLEVE	Boiling Liquid Expanding Vapor Explosion
CAA	Clean Air Act
CAAA	Clean Air Act Amendments
CAER	Community Awareness and Emergency Response
CAMEO	Computer Aided Management of Emergency Operations
CAS	Chemical Abstract Service
CCDF	Complementary Cumulative Distribution Function
CCPS	Center for Chemical Process Safety
CDO	Criteria for Design and Operations
CEPP	Chemical Emergency Preparedness Plan
CEPS	Center for Chemical Process Safety
CERCLA	Comprehensive Environmental Response Compensation Liability Act
CFR	Code of Federal Regulations
CMA	Chemical Manufacturer's Association
DOT	Department of Transportation
EAL	Emergency Action Level
EBS	Emergency Broadcast System
EG	Ethylene Glycol
EHS	Extremely (in N.J., Extraordinarily) Hazardous Substance
EO	Ethylene Oxide
EOC	Emergency Operations Center
EOL	Electrical One-Line Diagrams
EOP	Emergency Operations Plan

EP	Extraction Procedure
EPA	U.S. Environmental Protection Agency
EPCRA	Emergency Planning and Community Right-to-Know Act (SARA Title III)
EPZ	Emergency Planning Zone
ERPG	Emergency Response Planning Guideline
ESD	Emergency Shut-down
FAA	Federal Aviation Administration
FCV	Flow Control Valve
FEMA	Federal Emergency Management Agency
FMEA	Failure Modes and Effects Analysis
FTA	Fault Tree Analysis
HAZOP	Hazard and Operability Study
HAZWOPER	Hazardous Waste Operations and Emergency Response (OSHA 1910.120)
HEL	Higher Explosive Limit
HEP	Human Error Probabilities
HF	Hydrofluoric Acid
HFE	Human Factors Engineering
HMAC	Hazardous Materials Advisory Council
HRA	Human Reliability Analysis
ICS	Incident Command System
IDLH	Immediately Dangerous to Life and Health
IRC	Incident Response Commander
LC50	Median Lethal Concentration
LCLO	Lethal Concentration Low
LD50	Median Lethal Dose
LDLO	Lethal Dose Low
LEL	Lower Explosive Limit
LEPC	Local Emergency Planning Committee
LOC	Level of Concern
LPG	Liquefied Petroleum Gas
MEA	Monoethanol Amine
MHI	Material Hazard Index
MIC	Methyl Isocyanate
MOV	Motor Operated Valve
MSDS	Material Safety Data Sheet
MTBF	Mean Time Between Failures
MTTR	Mean Time To Repair
NFPA	National Fire Protection Association
NIOSH	National Institute of Occupational Safety and Health
NOAA	National Oceanic and Atmospheric Administration
NRT	National Response Team

NRT-1	Hazardous Materials Planning Guide
OPA 90	Oil Pollution Act of 1990
ORC	Organization Research Counselors, Inc.
ORI	Organizational Resources Incorporated
OSHA	Occupational Safety and Hazard Act
OSHA	Occupational Safety and Health Administration
P&ID	Piping and Instrumentation Diagram
PCB's	Polchlorinated Biphenyls
PCV	Pressure Control Valve
PEL	Permissible Exposure Limit
PFD	Process Flow Diagrams
PHA	Preliminary Hazards Analysis
PPE	Personal Protective Equipment
PSF	Performance Shaping Factor
PSV	Pressure Sensing Valve
RCRA	Resource Conservation and Recovery Act
REL	Recommended Exposure Limit
RMP	Risk Management Program
RMPP	Risk Management and Prevention Program (California)
RQ	Registration Quantity
RTECS	Registry of Toxic Effects of Chemical Substances (from NIOSH)
RV	Relief Valve
RVP	Reid Vapor Pressure
SARA	Superfund Amendments Reauthorization Act of 1986
SCBA	Self-Contained Breathing Apparatus
SCP	Standards Completion Program
SERC	State Emergency Response Commission
SHI	Substance Hazard Index
SIC	Standard Industrial Classification
SOP	Standard Operating Procedure
TCPA	Toxic Catastrophe Prevention Act (New Jersey)
THERP	Technique for Human Error Rate Prediction
TLV	Threshold Limit Value
TPQ	Threshold Planning Quantity
TWA	Time Weighted Average
UCVE	Unconfined Vapor Cloud Explosion
UEL	Upper Explosive Limit
UNEP	United Nations Environmental Programme
UPS	Uninterruptible Power Supply
VP	Vapor Profile
VS	Vapor Sample

Index

Abbreviations, rule for, 140
Accident evaluation, 279–80
Accident explosion and fire analysis, 196–97
Accident investigation, 9–10
  objectives of, 9
Accident investigation report, 10
Acetic anhydride, 28–29
Acronyms, 353–55
Active failure, definition, 139
Acute exposures, 213
Acute toxicity concentration (ATC), 26, 212
Acutely hazardous materials, 104
  risk management legislation, 346–49
Administrative control, 5
Alarm systems, for industrial facilities, 272–73
Alert emergency level, 279
Alternative risk financing techniques, 311–14
  banking plans, 313
  captives, 313
  cash-flow plans, 311–12
  fronting plans, 312–13
  risk retention groups and risk purchasing groups, 313–14
American Conference of Governmental Industrial Hygienists Threshold, 23
American Industrial Hygiene Association (AIHA), 213

Emergency Response Planning Guidelines of, 26
American Institute of Chemical Engineers (AIChE) Center for Chemical Process Safety (CCPS), 3, 102, 162
American Petroleum Institute (API), 12
  Management of Process Hazards of, 352
Ammonia, boiling point of, 191
Ammonia storage accident, 189–92
"And/Or" logic, 128–30
"And/Or" modeling, 129–30
  logic device example, 130f
Army Technical Manual, 207
Artificial intelligence, (AI), 317
Assessment of health effects from chemical releases, 209–19
  assessing magnitude of potential release, 211–12
  characterizing health risks, 214–15
  chemical characterization, 210–11
  determining toxicity, 212–13
  estimating exposure, 213–14
  prevention process, 216–17
  prioritization process, 215–16
  source of information, 217–19
ATC (Acute toxicity concentration), 26, 212
Atmospheric turbulence, 179
Audits, 10–11
  administration of, 10
  audit program requirements, 10

Banking plans, 313
Basic events, in fault tree diagram, 134
Bayesian methods, 334
Bernoulli flow equation, 172–74, 189
Bhopal tragedy, 263, 308, 336, 343–44, 352
Blackboard system, 325–26
BLEVE. *See* Boiling liquid expanding vapor explosion
Boiling liquid expanding vapor explosion (BLEVE), 196, 197, 198–99
  thermal radiation of, 198
Boiling point, 210
Boolean algebra, 133, 143–49
  cancellation/absorption laws, 146–47
  complementation laws of, 147, 148f
  computers and, 146
  definition, 143
  fault tree analysis and, 143–44
  fault tree equivalence and, 145f
  rearrangement laws, 146
  results in reduction process, 148
  symbols in, 144t, 144–145
  variables with one of two values, 144
Bouyancy-induced turbulence, 179
Burning rate, 201
Business Income Coverage Form, 309
Business interruption, 309–10

CAAA. *See* Clean Air Act Amendments
California Risk Management Prevention Plan, 84
CAMEO Response Information Data Sheet, 23
Captives, 313
Cash-flow plans, 311–12
Catastrophic failures, chemical releases from, 197
Catastrophic ruptures, 172
Cause-consequence analysis, 232

Cause-consequence diagram, 232, 233
  sample, 234f
Center for Chemical Process Safety (CCPS) of American Institute of Chemical Engineers (AIChE), 162, 352
Centerline maximum volume concentration, 202–3
Checklist reviews, 30–47
  advantages and disadvantages of, 32–33
  description of methodology, 30–31
  example checklists, 33–47
  methodology implementation, 32
  objective of checklist, 31
  results, 32
  supporting documentation, 31
Chemical(s)
  flammability of, 210–11
  physical state of, 210
Chemical Emergency Preparedness Program of Environmental Protection Agency (EPA), 345
Chemical hazard screening, 23
Chemical inventory identification process, 24
Chemical Manufacturer's Association (CMA), 3, 8, 264
  Community Awareness and Emergency Response (CAER) program, 345, 352
Chemical plants, expert systems for, 323–24
Chemical plume dispersion analysis, 167–94
  dispersion modeling, 178–88
  examples of possible release mechanisms, 170f
  sample problems, 189–94
  source definition, 168–71
  specific models, 189
Chemical screening process flow diagram, 25f
Chemical releases
  assessing magnitude of potential releases, 211–12

Chemical releases (*continued*)
  assessment process for health and
    impact of, 209–15
  determining, 197
  extent of risk from, 209
  rate of, 211
Chernobyl, 336
Chlorine cylinder accident, 192–94
Chronic exposures, 213
Clean Air Act Amendments (CAAA),
  11, 346, 350
Command fault, definition, 139
Common cause failures, 151–53, 153f
  fault tree example, 154t, 155
Community Awareness and
    Emergency Response (CAER),
    264
Compensating balance plans, 312
Complementary cumulative
    distribution functions (CCDF),
    228
"Complete the Gate" rule, 141
Component failure data, 224
Component identification, 98
Computer Based Instruction, 258
Computer Based Tutorial, 258
Computer techniques, 316–41
  constraint-based reasoning, 340–41
  in crisis management, 336–40
  defining expert systems, 317–20
  expert systems, 316
  system methodology, 320–24
Consequence analysis, 3–4
Constraint-based reasoning, 340–41
Constraint-based system, 340–41
  object of, 340
Construction checklists, 32
Continuation symbol, in fault tree
    diagram, 135
Continuation training, 258–59
Continuous gaseous release, 174
Continuous ground-level plume
    release, 184–85
Continuous liquid release, 172–74
Control (or treatment) management,
  4–5

Credit factor, 19
Crisis management, 336–40
  characteristics of, suitable for
    expert systems, 336
  expert systems and, 316
  spill management, 337–40
Critical components/events, 241
  in fault tree, 161
  risk reduction measures for, 162t
Criticality ranking system, 85–87
Criticality rating, example, 111
Cut sets, 144

*Dangerous Properties of Industrial
    Materials,* 218
Data
  collection and management of,
    222–23
  manipulation of, 225
Data applications
  background, 221–22
  data collection and management,
    222–23
  data manipulation, 225
  typical sources of generic data,
    224–25
Data base, 318
DEGADIS model, 180, 189
Demand rate data, 224
Dempster-Shafer theory of evidence,
  335
Design life of plant, five stages of,
  xiv–xv
Directors and officers (D & O)
    liability, 310
Discrete probability distributions
    (DPD), 228
Dispersion modeling, 167, 178–88
  comparison of concentration/
    emission rates, 188f
  cumulative frequency distributions,
    187f
  heavy gas dispersion, 180–82
  jet release dispersion, 179–80
  model performance and uncertainty,
    185–88

neutrally bouyant gases, 182–85
thermodynamic aspects of typical
  hazardous material release,
  182f
Dispersion models
  examples of, 190–91, 191t
  performance of, 186–88
Domain experts, 322, 334
DOT. *See* U.S. Department of
  Transportation
Dow Chemical Company, 15
Dow Fire and Explosion Index, 15–19
  basis procedure for index
    calculation, 16
  concept of, 15
  credit factor, 19
  first edition of, 15
  general process hazards, 17
  purpose of, 16
  sample index, 18f
  special process hazards, 17
  unit damage factor, 19
  unit hazard factor, 17
Dow/Mond, 3
Dupont Corporation, 316

Editors and contributors, biographic
  profiles of, xvii–xxii
Electrical one-line diagram (EOL), 91
Emergency/emergencies
  facilities, 297–98
  three-level classification system of,
    279–80
Emergency Action Levels (EALs),
  279
Emergency Director, 279, 282, 284,
  287
Emergency functions, 285–88
Emergency operations center (EOC),
  268–70, 270f, 271f, 273, 297, 298
Emergency organization, tasks to be
  carried out by, 8
Emergency planning, 8–9
Emergency Planning and Community
  Right-to-Know Act (EPCRA), 8,
  27, 263, 345

Emergency planning at local level,
  291–301
  emergency management, 291–92
  leadership commitment, 292–93
  plan and procedures, 299–301
  planning process, 294
  planning team, 293–94
  planning team tasks, 294–96
  resources, 296–99
Emergency planning for industrial
  facilities, 265–78
  communication equipment and
    alarm systems, 272–74
  emergency operations centers
    (EOC), 268–70
  fax machines, 274
  firefighting facilities, equipment, and
    supplies, 275–76
  media center, 271–72
  medical facilities, equipment, and
    supplies, 277
  meteorological equipment, 277–78
  monitoring systems, 277
  personal protective equipment
    (PPE), 274–75
  resource assessment, 267f
  resources, 266–68
  security and access control
    equipment, 278
  spill and vapor release control
    equipment, 276–77
  transportation equipment, 278
Emergency preparedness, 263–303
  accident evaluation, 279–80
  emergency planning at local level,
    291–301
  emergency planning for industrial
    facilities, 265–78
  emergency response action, 288–89
  emergency response organization,
    280–88
  public information, 301–3
  training, exercises, and plan
    maintenance, 289–91
Emergency response actions,
  288–89

Emergency response implementing
procedures, 288–89
Emergency response organization,
280–88
Emergency Director, 282, 284, 287
emergency functions, 285–88
emergency responsibility matrix,
281f
full emergency response
organization, 282–84
Incident Commander, 282, 285, 287
initial response organization, 281–82
Response Operations coordinator,
282, 284–85
Emergency Response Planning
Guidelines, 214
"Emergency Response to Hazardous
Substance Releases," 265
Emergency response training, 9
Engineering control, 5
Entrainment assumptions, 181
Environmental factors, 49–50, 56
Environmental and Field Survey, 286
Environmental impairment liability,
308
Environmental Protection Agency.
See U.S. Environmental
Protection Agency (EPA)
Evaporating pool, 177f
Evaporation rate, 176–77, 191, 192, 201
EVC (equilibrium vapor
concentration), 26
Event tree analysis (ETA)
use of, 128
when to use and when not to use,
165–66
Event trees, 156–61
specific event sequence, 157–59
typical example, 158f
Existing facilities
safety audit example, 61–65
safety audit for, 58–59
sample electrical one-line diagram
(EOL) checklist, 67f
sample P&ID checklist, 64–65

sample process flow diagram (PFD)
checklist, 66f
Expert system methodology, 320–24
chemical and process plants, 323–24
diagnosis of problems, 322–23
expert system shells, 320–21
system development, 321–22
Expert systems, 316
data base, 318
defining, 317–20
development of, 319–20, 320f,
325–27
elements in, 318
explanatory facility, 318
inference engine, 318
interpretation of data in, 327
knowledge base, 318
operations- and control-related
activities well suited to,
324–325
in plant control problems, 325
real-time expert systems, 324–28
for risk assessment, 328–35
rule base, 318–19
user interacting with, 319f
Explanatory facility, 318
Explosion and fire analysis, 196–207
analysis methodology, 196–98
combustible chemicals in, 196–97
effect of explosive pressure on
human health, 207t
effect of fires and explosions on
health and the plant, 207
effect of heat flux on human health
and the plant, 207t
fire and explosion characteristics,
198–206
generation of missiles, 198
pool fire study for n-Heptane, 203f
propane tank failure, 200f
release flow rates, 197
vapor, liquid, or solid, 197–98
Explosions, 4
Exposures
boiler and machinery, 309

in chemical process industry,
307–11
directors and officers (D & O)
liability, 310
environmental impairment liability,
308
estimating, 213–14
fidelity, 311
general liability, 310–11
inland marine, 311
product liability, 307–8
property, 308–9
vehicle liability, 310
workers' compensation, 309
Extraordinarily hazardous substance
(EHS), 26, 27t
Extremely hazardous substance
(EHS), 27, 291
Extremely Hazardous Substances List
and Threshold Planning
Quantities, 23
Extremely Hazardous Substances
Risk Management Act, of
Delaware, 350
Extrinsic safety, xv

Facilities, review of, 50
Failure
probability of, 222
as term, 139
Failure effect, definition, 139
Failure mechanism, definition, 139
Failure mode, definition, 139
Failure modes and effects analysis
(FMEA), 2, 3, 91–100
advantages and disadvantages, 100
data sheet design, 97–99
definition, 91
documents and drawings required,
91–95
level of detail, 97
methods for documenting study, 100
review of documentation, 99–100
sample FMEA data sheet, 92–93t
selection of team members, 95–96

study guidelines, 96–99
system definition, 96–97
Failure rate data, 224
Familiarization and qualitative
assessment, 244–45
Farmer risk assessment curve, 235,
236f
Fault, as term, 139
Fault category, as term, 139
Fault identification (The "Be Precise
Rule"), 140
Fault tree(s)
common cause failures, 151–56
creating "complement" of, 159
example: event sequence, 160f
example: single pump system and
fault tree, 143
identifying potential common cause
failures, 153–54
Fault tree analysis (FTA), 2, 127–28
Boolean algebra and, 143–44
cardinal roles of fault tree
quantification, 162–63
computer program for, 164
definition/description, 127–28
history of, 127–28
qualitative results, 161–62
quantitative results, 162–64
reducing fault tree, 163
sample computer codes for, 165t
when to use and when not to use,
165–66
Fault tree construction
guidelines for, 139–43
no "Gate-to-Gate" rule, 142f
simple lighting circuit, 136f
single system and fault tree, 137f
Fault tree diagram
symbology in, 134, 135f
transfer in, 135
Fault tree equivalence, 149–50
Fault Tree Handbook, 134, 140, 164
Fault tree reduction, 148, 149–50
Fax machines, in industrial facilities,
274

Federal Clean Air Act Amendments (CAAA), 11
Federal Emergency Management Agency (FEMA), 294, 299
Federal legislation and regulations, 345–46, 350–52
Federal Occupational Safety and Health Administration (OSHA). *See* Occupational Safety and Health Administration (OSHA)
Feedstock review, 48
Feedstocks, 48, 53–54
FID (flame ionization detector), 210
Fidelity, 311
Fire, 4. *See also* Explosion and fire analysis
Fire and explosion index. *See* Dow Fire and Explosion Index
Fireball, 198–99
Firefighting facilities, equipment, and supplies, 275–76
Flammability, 24
Flashing liquids, 175
Flash point, 211, 216
Freezing point, 210
Frequency, 111–12
Fronting plans, 313
Functional annexes, 300

"Gate-to-Gate" rule, 141
Gaussian dispersion model, 180–87, 184f
  accuracy of, 186
  assumptions in, 185–86
General emergency, 279–80
General liability, 310–11
General operations checklist, 35–38
General process hazards, 17
Generic data, 222
  typical sources of, 224–25
Genium Publishing Corporation, 218
Guidance for the Preparation of a Risk Management and Prevention Program, 213
Guidelines for Chemical Process Quantitative Risk Analysis, 164

The Handbook of Artificial Intelligence, 317
Handbook of Chemistry and Physics, 202
Handbook of Toxic and Hazardous Chemicals and Carcinogens, 218
Hazard, definition, 102
Hazard and operability study (HAZOP), 2, 3, 101–25
  aim of, 102
  checklist, 45–47
  definition/description, 101, 102
  discussion of previous checklist items, 124–25
  ethylene glycol production HAZOP example, 110t
  example problem, 109–14
  formal report, 122–23
  guide word approach, 102, 103–5
  hazard and operability study action report, 106f, 116–17, 117f, 118–19, 119f, 120–21, 121f
  importance of documentation in, 103
  interim report, 115–22
  keys to success of, 109
  level of detail, 97
  maintaining HAZOP record, 115
  mini-report, 115
  participants in, 105
  performing, 102–3
  plant walk-through, 114–15
  preliminary hazard identification, 109–11
  preplanning issues, 107
  procedural review in, 104
  recording team proceedings, 114–15
  risk matrix, 112
  role of, 103
  simple presentation, 109
  study logistics, 107–8
  summary of critical identified hazards, 123f
  typical agenda, 108–15
  uses of, 101–2
"Hazard assessment," 4

Hazard Communication Act, 217
Hazard Communication Standard, 263
Hazardous chemicals
    examples of possible release
        mechanisms, 170f
    practical processes in emission and
        dispersion of, 167–68
    release pathways of, 168–69
    toxic or explosive vapor clouds
        from, 169
Hazardous components, 4
Hazardous compounds, 4
Hazardous material inventories, 84
Hazardous material models, 186
Hazardous Waste Operations and
    Emergency Response
    (HAZWOPER), 264
Hazards analysis, 294
Hazards identification, 2–3, 294
Hazard-specific appendices, 300
HAZMAT, 286
HAZOP. *See* Hazard and operability
    study
Heat balance equation, 175
Heat flux, due to convection, 178
Heavy gas dispersion, 180–82
    rate of spread of cloud, 181
Higher explosive limit (HEL), 202–3
Horizontal dispersion coefficient,
    202
Hotlines, in industrial facilities, 274
HRA. *See* Human reliability analysis
Human error, 336
Human error performance, sample
    data, 242–43, 243f
Human error probabilities (HEPs),
    241
Human factors engineering (HFE),
    238
Human reliability analysis (HRA),
    238–52
    application example, 249–52
    documents or input information
        required for, 241–43
    event trees, 245–49, 250f
    familiarization, 239

    guideline for performance of,
        244–49
    guidelines for selection of HRA
        team leader and members,
        243–44
    history and background, 238
    HRA process, 239–41
    job task analysis sample, 246t
    overview of, 240f
Human reliability analysis process,
    239–41
    incorporation of data, 241
    qualitative assessment, 239
    quantitative assessment, 241

Incident Commander, 268, 282, 285,
    287
Incurred loss retro, 312
Industrial relations. *See* Training for
    industrial relations
Industry standards, 12
Inference engines, 318, 322, 323
Inherently safe, xiv–xv
Initial Notification and Public Alert
    and Warning functional annexes,
    302
Initial or startup operation
    checklists, 32
Initial training, 258
Inland marine, 311
Immediate causes (The "Think Small
    Rule"), 140
Immediately dangerous to life and
    health (IDLH), 212, 295
Imperial Chemical Industries (ICI),
    102
    Mond Division of, 19
Inspection checklist, 41–43
Insurance
    benefits of, 306
    self-insurance versus, 305–7
Insurance premium, components of,
    305–6
Insured cash flow plans, 312
Interconnecting systems, 49

Intermediate events, in fault tree
diagram, 134
Involuntary risks, 301
Ionization potential, 210

Jet release dispersion, 179–80

Kemeny Commission Report on the
Three Mile Island Unit 2
accidents, 238
Knowledge base, 318

Legislation and regulation, 343–52
additional federal actions, 350–52
federal action, 345–46
industry action, 352
required program elements, 351t
state initiatives, 346–50
Lethal concentration low (LCLO), 212
Lethal dose low (LDLO), 212
Level of concern (LOC), 214
Limit Values and Biological Exposure
Indices, 23
Linked fault trees and success trees,
159–61
Liquid pool evaporation, 175–78
LISP computer language, 320
Local Emergency Planning
Committees (LEPCs), 291
Logic gates, in fault tree diagrams,
135
Logic modeling, 128
fault tree diagram in, 134, 135
graphical symbols for, 134–38
Logic operators, 130
Lower explosive limit (LEL), 202,
204–5, 211

Magnitude, as term, 211
Maintenance checklist, 38–41
Maintenance procedures, 6
Maintenance procedures
documents, 83
Major components review, 48–49
Major process components, 48–49, 54

Management of Process Hazards,
23, 26
Material Hazard Index (MHI),
formula for, 216
Material Safety Data Sheet (MSDS),
217, 287
A Mathematical Theory of Evidence,
335
McCarran-Ferguson Act, 313, 314
Mechanical failures, 91
Mechanical turbulence, 179
Median lethal concentration (LC50),
212
Median lethal dose (LD50), 212
Medical facilities, equipment, and
supplies, 277
Melting point, 210
Meteorological equipment, in
industrial facilities, 277–78
Methyl isocyanate (MIC), 343–44
Microcomputers, 321
Mitigative systems, 51
Mond Division of Imperial Chemical
Industries (ICI), 19
Mond Fire, Explosion, and Toxicity
Index, 19–22
loss control credit factors, 20f
steps for performing, 22
as two-part calculation, 22
uncertainty and, 185–88
unit analysis summary, 21f
Monte Carlo simulation, 226
Multihazard emergency operations
plan (EOP), 299
MYCIN program, 317, 322, 334

National Academy of Sciences, 317
National Aeronautical and Space
Administration, 198
National Chemical Safety Board,
350
National Fire Protection Academy
(NFPA), 216
National Fire Protection Association
(NFPA), 23, 276

hazard identification system of, 27–28

National Institute of Occupational Safety and Health (NIOSH), 212, 295

Pocket Guide of Chemical Hazards, 218

National Oceanic and Atmospheric Administration, 23

National Response Team's Hazardous Materials Emergency Planning Guide (NRT-1), 296

New Jersey Department of Environmental Protection (NJDEP), 347, 350

New Jersey Toxic Catastrophe Prevention Act (TCPA). *See* Toxic Catastrophe Prevention Act

The "No miracles rule," 140–41

Numerical simulation models, 327

Occupational Safety and Health Administration (OSHA), 11–12, 26, 84, 213, 263, 272, 291, 346, 350

Off-line interactive expert systems, 324

Off-site liaison, 287

Operability, definition, 102

Operating procedures, 6

Operating procedures documents, 83

Operations checklists, 32

Operator job descriptions, 83–84

Organization Resources Counselors, Inc. (ORC), 350

OSHA. *See* Occupational Safety and Health Administration

Paid loss "Retro," 311–12

Pasquill-Gifford curves, 185

Passive failure, 139

PC-based expert systems, 321

PCs. *See* Personal computers

Performance shaping factors (PSFs), 239

Permissible exposure limit (PEL), 212

Personal computers (PCs), 321

Personal protective equipment (PPE), 274–75

Personnel, 296

Phosgene, 210

PID (photoionization detector), 210

Piping and instrumentation diagrams (P&IDs), 84, 91

Piping and instrumentation drawings (P&IDs), 3

Plan integration, 301

Planning team tasks, 294–96
  hazards analysis, 294
  hazards identification, 294
  risk analysis, 295–96
  vulnerability analysis, 294–95

Plant control problems, 325

Plant design checklists, 31

Plant management, safety audits and, 60

Plant operation checklist, 31

Plant organization and administration checklist, 33–35

Plant reentry, 289

Plant risk assessment, 19–22

Plant specific data, 222

Pool fires, 199–202

Postemergency activities, 289

Preliminary hazards analysis, 48–56
  advantages and disadvantages of, 52
  description of methodology of, 48–51
  example problem, 52–56
  methodology implementation, 51–52
  preliminary hazards analysis table, 55t
  purpose of, 51
  results, 52
  supporting documentation, 51

Premium payment plans, 312

Primary events, in fault tree diagram, 134

Primary fault, definition, 139

Probability, definition, 221–22

Process chemistry description document, 83
Process flow diagrams (PFDs), 84
Process plants, expert systems for, 323–24
Process Safety Management of Highly Hazardous Chemicals, 26
Proposed operations, 50–56
Public address systems, in industrial facilities, 272–73
Public information, 301–3
  emergency public information, 302–3
  public education, 301–2

Quantified risk assessment, 221–37
  application example, 232–37
  benefit of, 232
  brief review of data applications, 221–25
  communicating quantified risk assessment results, 228–32
  example risk calculation, 235t
  uncertainty and sensitivity analysis, 225–28
Quantifed risk assessment results, 228–32
  acceptability of risk, 231–32
  objective of risk assessment, 228
  relative risks, 228
  risk contours, 230–31

Radios, in industrial facilities, 274
Rankings, 113t
Reactivity, 24
Reactivity data, 210
Reactor Safety Study (WASH-1400), 228
Read and Sign Assignment instruction technique, 257–58
Real-time expert systems, 324–28
  benefit of, 324
  blackboard system, 325–26
  development and implementation of, 324
  types of plants for, 327–28

"Recommendations for Process Hazards Management of Substances with Catastrophic Potential," 350
Recovery phase, 300
Red Cross, 298
Reference sources, 23
Refresher training courses, 7
Resource Conservation and Recovery Act (RCRA), 84
Resources, 296–99
  equipment, 298
  facilities, 297–98
  personnel, 296
Respirators, in industrial facilities, 275
Response Operations Coordinator, 282, 284
Restoration of services, 289
Reverse-order confirmation, 31
Right-to-Know Law, 217, 263
Ring-down systems, in industrial facilities, 274
Risk
  acceptability of, 231–32
  formal definition of, x
  graphical presentation of, 230f
Risk analysis, 295–96
Risk assessment, 328–35
  confidence factors, 334–35
  expert systems and, 316
  objective of, 228
  probabilistic risk assessment, 330–34
Risk contours, 230–31
Risk financing, 305–14
  alternative risk financing techniques, 311–14
  exposures in chemical process industry, 307–11
  insurance versus self-insurance, 305–7
Risk Management and Prevention Program (RMPP) of California, 264
Risk management programs, 1–13
  accident investigation, 9–10

administrative control, 5
audits, 10–11
change control in, 5–6
consequence analysis, 3–4
control or treatment management,
    4–5
current and pending regulation,
    11–12
elements of, ix
emergency planning, 8–9
hazards identification, 2–3
key elements in, 2
procedures, 5–6
training, 6–7
Risk matrix, 112
Risk perceptions
acceptable versus unacceptable,
    xi–xiii
"gradual" versus "sudden," xii
natural versus artificially
    constructed risk, xii
of public compared with experts, xi
risks beneficial to individual, xiii
"usual" versus "unusual," xii
voluntary versus involuntary risks,
    xii–xiii
Risk purchasing groups, 313–14
Risk rankings, 112, 113
Risk Retention Act, 313
Risk retention groups, 313–14
Rule base, 318–19
Bule-based expert systems, 322–23
Rule-Based Expert Systems: The
    MYCIN Experiments of the
    Stanford Heuristic Programming
    Project, 334

Safeguards review, 51
Safety audits, 57–74
advantages and disadvantages,
    60–61
definition/description, 57
description of methodology, 57–58
example audits, 61–73
methodology implementation, 59–60
periodic reviews, 57

purpose of, 60
reporting results, 60
suggested readings, 74
supporting documentation, 59
Safety checklist, 43–45
Safety components, 51
Safety engineering, xiii–xiv
Safety procedures, 6
Safety training, 7
SARA. *See* Superfund Amendment
    and Reauthorization Act (SARA)
    Title III
Screening analysis techniques, 15–29
Dow Fire and Explosion Index,
    15–19
general screening analysis
    techniques, 23–24
Mond Fire, Explosion, and Toxicity
    Index, 19–22
Secondary fault, definition, 139
Security and access control
    equipment, in industrial facilities,
    278
Self-contained breathing apparatus
    (SCBA), 275
Self-insurance, insurance versus,
    305–7
Sensitivity analysis. *See* Uncertainty
    and sensitivity analysis
Site emergency, 279
Solubility, definition, 210
Source term, 76
Special process hazards, 17
Spill management, 337–40
analysis, 339–40
characterization, 337–38
destination of discharge, 338
detection, 337
hazard assessment, 339
identification of source, 338
magnitude of release, 338
notification, 229
response determination, 339
Split payment plans, 312
Standard Industrial Classification
    (SIC) Codes, 20–39, 263

Standard operating procedures
(SOPs), 300
Standards Completion Program (SCP),
212
Standby or shutdown checklists, 32
State Emergency Response
Commission (SERC), 291
State regulations and legislation, 11,
346–350
Substance Hazard Index (SHI), 23,
24–26, 213, 216
expression for determining, 26
list of substances and corresponding
SHIs, 26f
Success criteria
defining, 134
four-pump example, 131f
success versus failure logic, 132f
Success logic and failure logic, 131
Success tree analysis, 128
Superfund Amendment and
Reauthorization Act (SARA) Title
III, 84, 217, 263, 264, 345, 346
Emergency Planning and
Community Right-to-Know Act
of 1986 (EPCRA), 291
Support Center for Regulatory Air
Models (SCRAM), 189
Support systems, 49, 54–56
Supporting documentation, 31
Surface area, calculating, 176
Sutton's equation, 201
System, definition, 97
System facilities, review of, 50
System failure, 130–33
System failure logic, 133–34
System of Hazard Identification, 23
System output signal, 133

*Technical Guidance for Hazards
Analysis: Emergency Planning for
Extremely Hazardous
Substances,* 211, 294, 295
Technical Specifications Advisor, 328,
330, 334

Technique for Human Error Rate
Prediction (THERP), 243
Telephones, in industrial facilities,
273–74
Thermal hydraulic analysis, 197
Three Mile Island incident, 263, 336
Threshold Planning Quantity (TPQ),
291
Time-weighted average-threshold limit
value (TWA-TLV), 212
Title III of Superfund Amendment
and Reauthorization Act (SARA),
See Superfund Amendment and
Reauthorization Act (SARA) Title
III
Top event, 140
Toxic Catastrophe Prevention Act
(TCPS), 26, 84, 87, 216, 264,
346–49
Toxic chemical accident model, 173f
Toxicity, 24
definition, 212
measuring, 212–13
Toxic leaks, 275
Toxic releases, 4
Training, 6–7
as term, 6
Training for industrial facilities,
253–61
documentation of training, 259–60
job analysis, 256
required training matrix, 260, 261t
training methods, 256–58
training program, 253–56
types of programs, 258–59
Training methods for industrial
facilities, 256–58
classroom lecture, 257
individual instruction, 257–58
laboratory practicum, 257
on-the-job training, 257
simulator training, 258
Training program for industrial
facilities, 253–56
evaluation, 255–56

facilities and equipment, 254
learning objectives, 255
lesson plans, 255
management support, 253
procedures, 254
scheduling, 254
Transportation equipment, in
    industrial facilities, 278
Treatment management, 4–5
Two-phase releases, 175

Uncertainty, definition, 226–27
Uncertainty analysis, use of, 225
Uncertainty and sensitivity analysis,
    225–28
background, 225
definition of uncertainty, 226–27
uncertainty analysis methods, 228
uses of, 227
Unconfined vapor cloud explosion
    (UVCE), 196, 197–98, 202–6
cloud model for, 204f
overpressure results for propane,
    206f
Undeveloped events, in fault tree
    diagram, 134
Union Carbide, 344
Unit hazard factor, 17
United States Armed Forces, 198
Upper explosive limit, 211
U.S. Department of Transportation
    (DOT), 294
U.S. Environmental Protection
    Agency (EPA), 8, 23, 27, 28t,
    291, 294, 346, 352
Chemical Emergency Preparedness
    Program of, 345
dispersion models, 189
Form R, 84
Support Center for Regulatory Air
    Models (SCRAM) of, 189
Utility and support systems, 49,
    54–56

Vapor
definition, 210
rate of release for, 211
Vapor density, definition, 210
Vaporization, 175–78
Vaporization rate, 191, 192
calculating, 178
Vehicle liability, 310
Venn diagram, 144, 147f, 148
Vertical dispersion coefficient, 202
Voluntary risks, 301
Vulnerability analysis, 294–95

Walk-throughs and facility tours, 258
WHAT-IF analysis, 2, 3, 75–90
advantages and disadvantages,
    88–89
basic characteristics of, 75–76
consequences recognized by, 77
criticality ranking system, 85–87
definition of impact area or receptor
    area for, 77
description of methodology, 75–83
documents, 78
example, 89–90
hazard analysis WHAT-IF analysis
    table, 80–82t
methodology implementation, 85
question formulation, 78–79
results, 85–88
sample table for results of, 86t
scope definition, 76
source and boundaries, 76–77
supporting documentation, 83–84
term selection, 77–78
use of, 75
WHAT-IF evaluation, 79–83
Wind socks, 278
Workbook Tutorial instruction
    technique, 257
Worker Right-to-Know, 263
Workers' compensation, 309
World Bank, 3